Diese Mitteilungen setzen eine von Erich Regener begründete Reihe fort, deren Hefte auf der vorletzten Seite genannt sind.

Bis Heft 19 wurden die Mitteilungen herausgegeben von J. Bartels und W. Dieminger. Von Heft 20 an zeichnen W. Dieminger, A. Ehmert und G. Pfotzer als Herausgeber.

Das Max-Planck-Institut für Aeronomie vereinigt zwei Institute, das Institut für Stratosphärenphysik und das Institut für Ionosphärenphysik.

Ein (S) oder (I) beim Titel deutet an, aus welchem Institut die Arbeit stammt.

Anschrift der beiden Institute:

3411 Lindau

X - R A Y   M E A S U R E M E N T S

I N   T H E   A U R O R A L   Z O N E

F R O M   J U L Y   T O   O C T O B E R   1 9 6 4

by

G. KREMSER, E. KEPPLER, A. BEWERSDORFF, K. H. SAEGER,
A. EHMERT, and G. PFOTZER

Institut für Stratosphärenphysik am
Max-Planck-Institut für Aeronomie
3411 Lindau/Harz, Germany

W. RIEDLER

Kiruna Geophysical Observatory
Kiruna C, Sweden

J. P. LEGRAND

Laboratoire de Physique Cosmique
Station de Chalais-Meudon
Meudon, France

ISBN 978-3-540-03365-3   ISBN 978-3-662-13448-1 (eBook)
DOI 10.1007/978-3-662-13448-1

- 3 -

Contents

I. Introduction.

In this report the data of balloon flights, carried out from July to October 1964 in Kiruna/Sweden are compiled. They were performed by the joint group of the Institut für Stratosphärenphysik am Max-Planck-Institut für Aeronomie, Lindau/Harz, Germany, of the Kiruna Geophysical Observatory, Kiruna, Sweden, and of the Laboratoire de Physique Cosmique, Meudon, France.

Herewith the sequence of flights started in 1960 by the Lindau and Kiruna group [PFOTZER et al. 1962a] and since 1963 executed in the frame of SPARMO is continued [PFOTZER et al. 1962b, PFOTZER et al. 1965].

In this campaign 53 balloons were launched, which carried scintillation counters, Geiger-Müller-tubes, and ionization chambers in different combinations.

In what follows one finds :

technical remarks concerning the balloons and the instruments (section II),

a description of the solar activity and its terrestrial effects during this balloon launching period (section III),

a list of the balloon launchings (section IV),

representations of the radiation measurements, remarks on the indications for launchings and some special comments (section V),

diagrams of the flights (section VI).

II. Technical remarks.

a) Balloons

As in the 1963 campaign [PFOTZER et al. 1965] tetrahedron type balloons of 5 000 m^3 were employed, which were manufactured by the " Centre de Lancement, Aire/l'Adour, du Centre National d'Études Spatiales, France ".

b) Instrumentation

The instruments used during this campaign differed only slightly from those launched in 1963. According to their composition they are denoted by : TESI, TESCI, TESIO, I.C.

TESI: A threefold coincidence telescope formed by three Geiger-Müller-counters, where also the counting rates of the topmost and the middle counters are telemetered to the ground (Victoreen 1 B 85 and 6306). The TESIS are manufactured by the factory Albin Sprenger K.G., St-Andreasberg/Harz [KEPPLER; 1964].

TESCI: A combination of a threefold coincidence telescope of Geiger-Müller-tubes (3 counters Victoreen 1 B 85) and a NaI (Tl)-scintillation counter. In some flights only the uppermost counter of the telescope was used to compare the counting rates of the Geiger-Müller-tube and the scintillation counter. The TESCIS were manufactured by the Institut für Stratosphärenphysik am Max-Planck-Institut für Aeronomie, Lindau/Harz, Germany [ROSSBERG and SPITZ to be published].

TESIO: Version of balloon instruments used until 1962, contained a vertically mounted Al-walled Geiger-Müller-counter, a telescope as in the TESCI and an ionization chamber (see I.C.). The TESIOS were manufactured by the Institut für Stratosphärenphysik am Max-Planck-Institut für Aeronomie, Lindau/Harz, Germany.

I.C.   Ionization chamber of the Neher-type.

Table 1 gives the characteristics of the various detectors.

<u>Table 1</u>

Characteristics of the various detectors.

Detector	Type	d/D cm	eff. length cm	absorb. layer of the wall mg/cm^2	geom. factor[*] cm^{2}ster	Energy-threshold (MeV) electrons	protons
Al-GM	1 B 85	1.9/-	7.0	30	75	0.16	3.8
Bi-GM	6306	1.9/-	7.0	135	75	0.4	5.5
Telescope	TESI	1.9/4.6	7.0	360	6.9	0.9	14
	TESCI	1.9/4.6	7.0	150	6.9	0.6	9
	TESIO	1.9/4.6	7.0	150	6.9	0.6	9
Ionization Chamber	NEHER	25/-	-	480	3080	1.1	16
Scintillator	TESCI	2.54/-	2.54	50	48	0.22	4.7

[*] Definition of the geometry factor G for isotropic radiations:

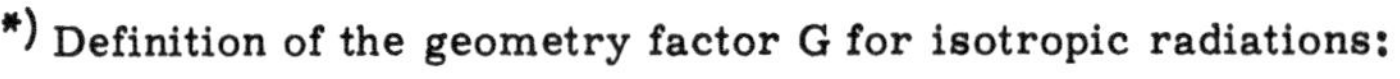

$$N = \epsilon \cdot G \cdot \Phi (E_{phot})$$

N = counting rate

ϵ = efficiency

G = geometry factor

$\Phi(E_{phot})$ = flux of monoenergetic photons/cm^2 sec sterad

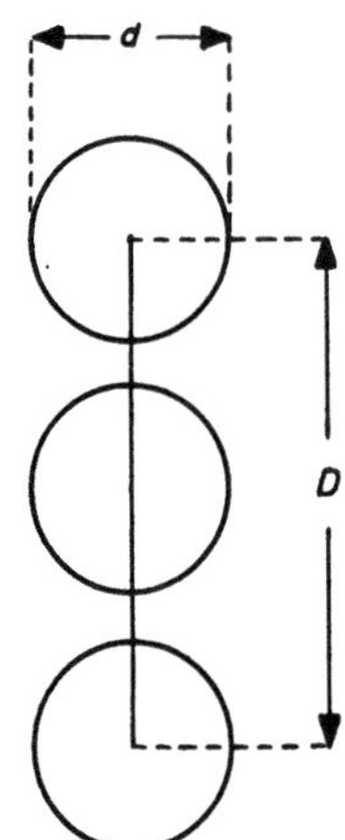

III. Solar activity and its terrestrial effects.

The balloon flights were performed during the solar rotations 1792 to 1795 of which following relevant features are represented in Figs. 1-4 :

the Zürich relative sunspot numbers (CRPL Boulder reports),
the position of plage regions on which flares occurred (CRPL Boulder reports),
the duration of balloon flights at Kiruna, marked by rectangular boxes , whereby the letter X above a box means that X-rays were measured during these flights,
the duration of balloon flights at other stations,
3-hourly planetary amplitudes ap (by courtesy of Geophysikalisches Institut, Univers. Göttingen),
hourly neutron counting rates at Deep-River (CRPL Boulder reports).

The solar activity during these rotations was very low due to the approach towards the minimum of the sunspot cycle. The relative sunspot number never reached 40 ; it was even zero during periods of ten and more days. Only a few flares of importance 1 and 2 occurred, none of importance 3.

In spite of this very low solar activity several weak or even medium geomagnetic storms occurred. Some of them began with an ssc and/or were accompanied by small Forbush effects (< 2%). However, none of the Forbush effects could be linked to a solar flare.

As the geomagnetic storms and the Forbush effects are manifestations of solar plasma clouds which are in some way also the source or the cause of the X-ray bursts, we have measured during our balloon flights, they shall now be described in more detail.

Solar rotation 1792 (July 2 to July 28, 1964) (Fig. 1)

Three weak geomagnetic storms occurred. Only the first one began with an ssc, on July 2 at 23.25 UT. No Forbush effect was observed on this day. This happened, however, in association with the third gradually commencing storm, on July 17. Flights K 1/64 to K 4/64 were started to cover this disturbance period.

As can be seen on Fig. 5 all storms are linked to more or less persistent M-regions passing the central meridian [BARTELS, 1932] . The first storm during this rotation is connected with an M-region, which recurred five times. The second one belongs to an M-region, which appeared in August 1962 and returned until October 1964. This region showed an outstanding recurrence tendency ; it recurred at least 30 times. The third storm occurred again during the recurrence of a less persistent M-region, which returned perhaps 8 times and which began 2 days earlier for the next three rotations.

Solar rotation 1793 (July 29 to August 24, 1964) (Fig. 2)

On August 5 a flare of importance 2 could even be observed in the visible light *) . But we cannot find terrestrial effects attributable to it. Again three geomagnetically disturbed periods belonging all to M-regions can be distinguished. In the second and third period weak geomagnetic storms occured, beginning with ssc's on August 4 at 01.30 UT and on August 11 at 00.55 UT respectively. Both were accompanied by small Forbush effects. Flights K 9/64 to K 12/64 were launched to cover the second disturbance period, flights K 13/64 to K 16/64 to cover the third one.

---

*) Ionosphären-Berichte, edited by Deutscher Wetterdienst, Hamburg, and by the Arbeitsgemeinschaft Ionosphäre, Darmstadt (Germany).

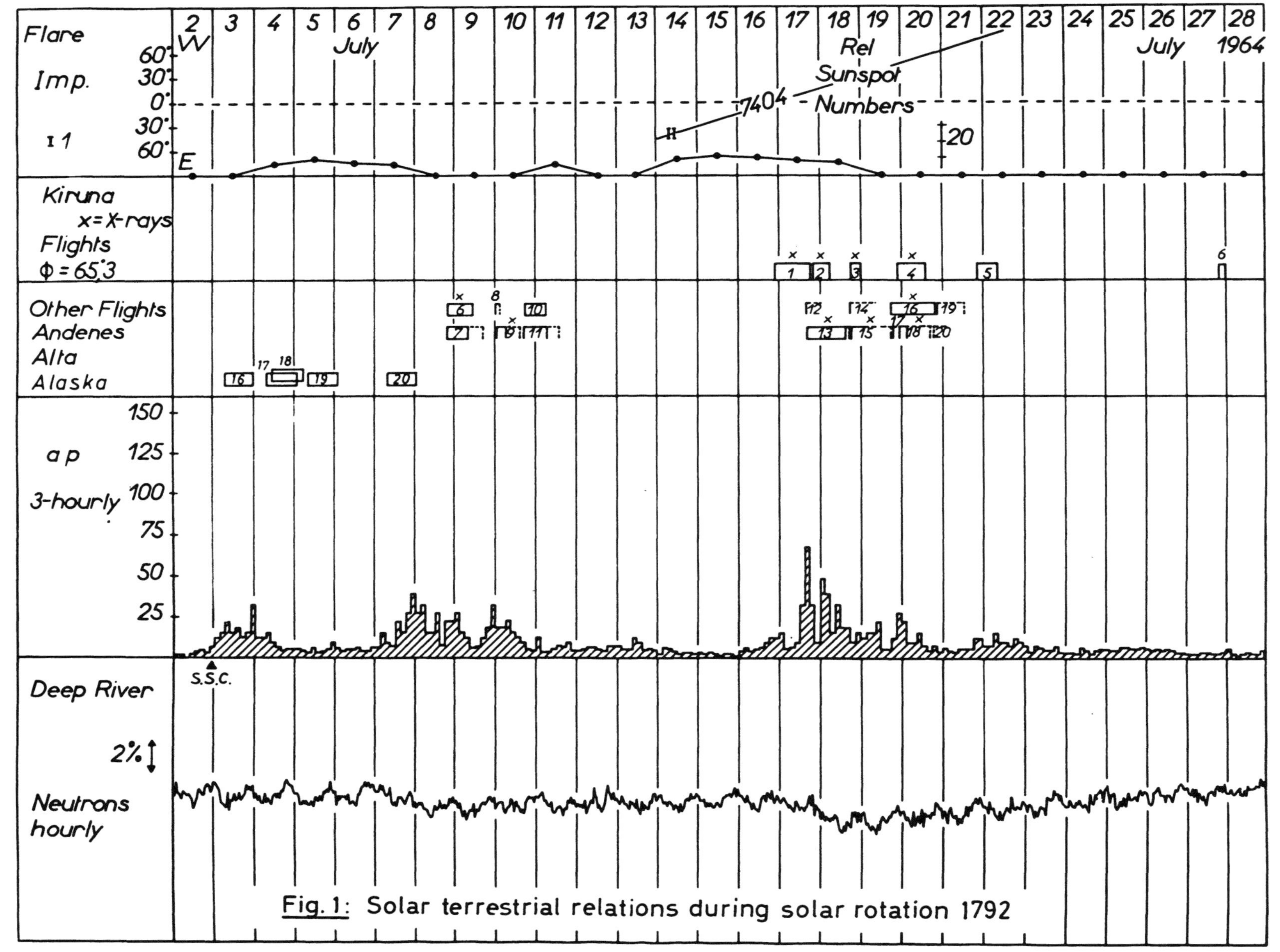

Fig. 1: Solar terrestrial relations during solar rotation 1792

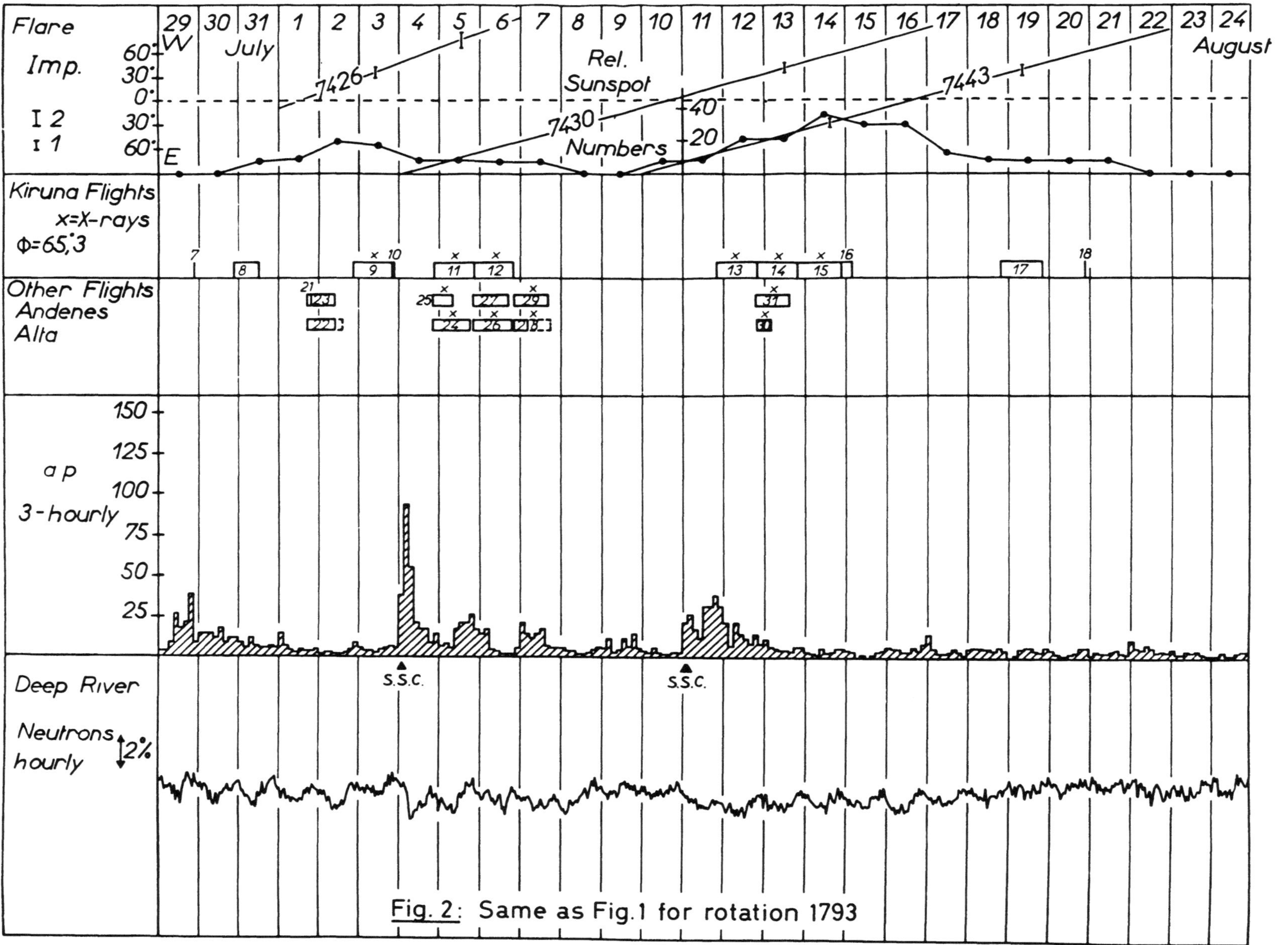

Fig. 2: Same as Fig.1 for rotation 1793

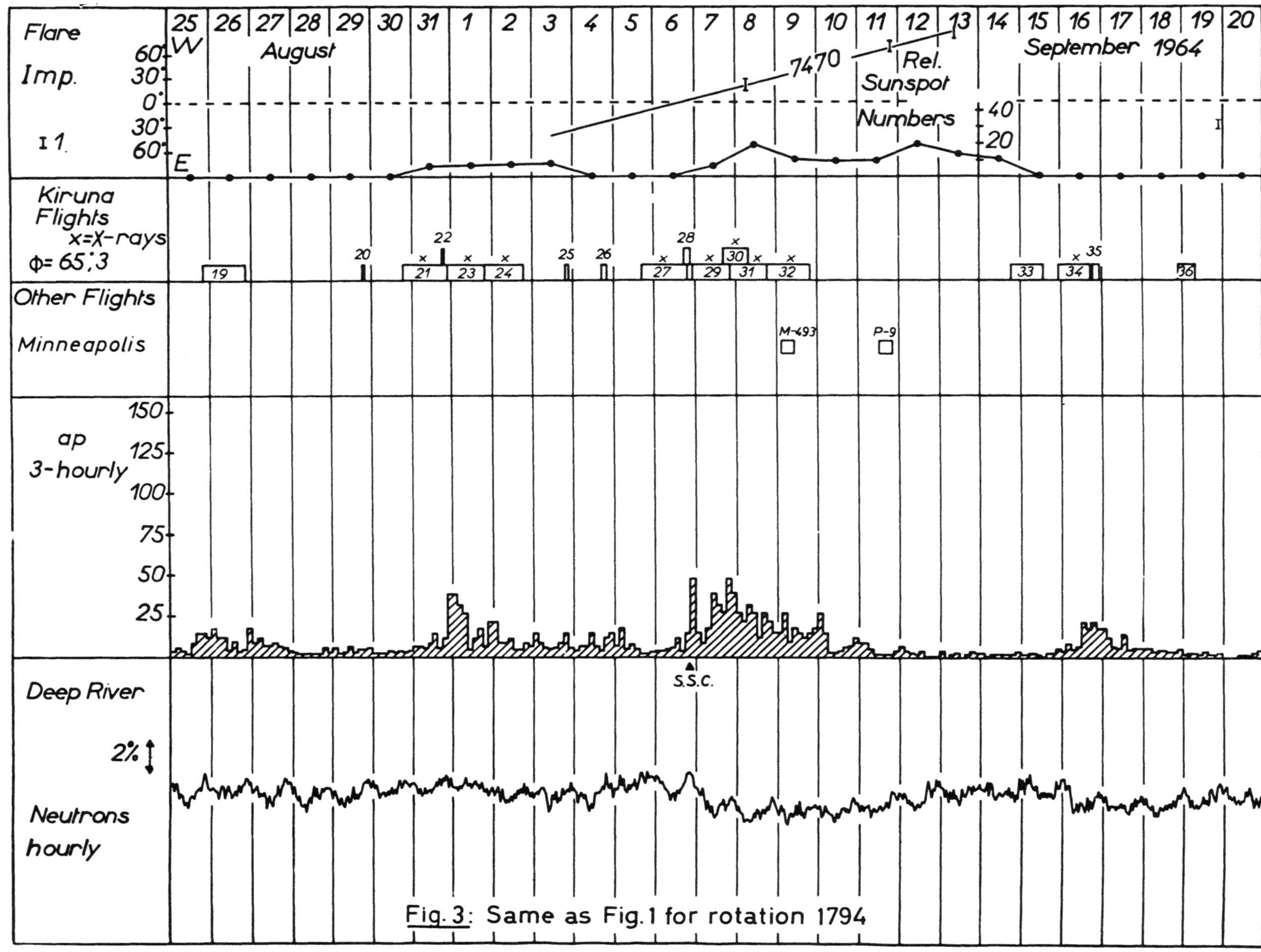

Fig. 3: Same as Fig. 1 for rotation 1794

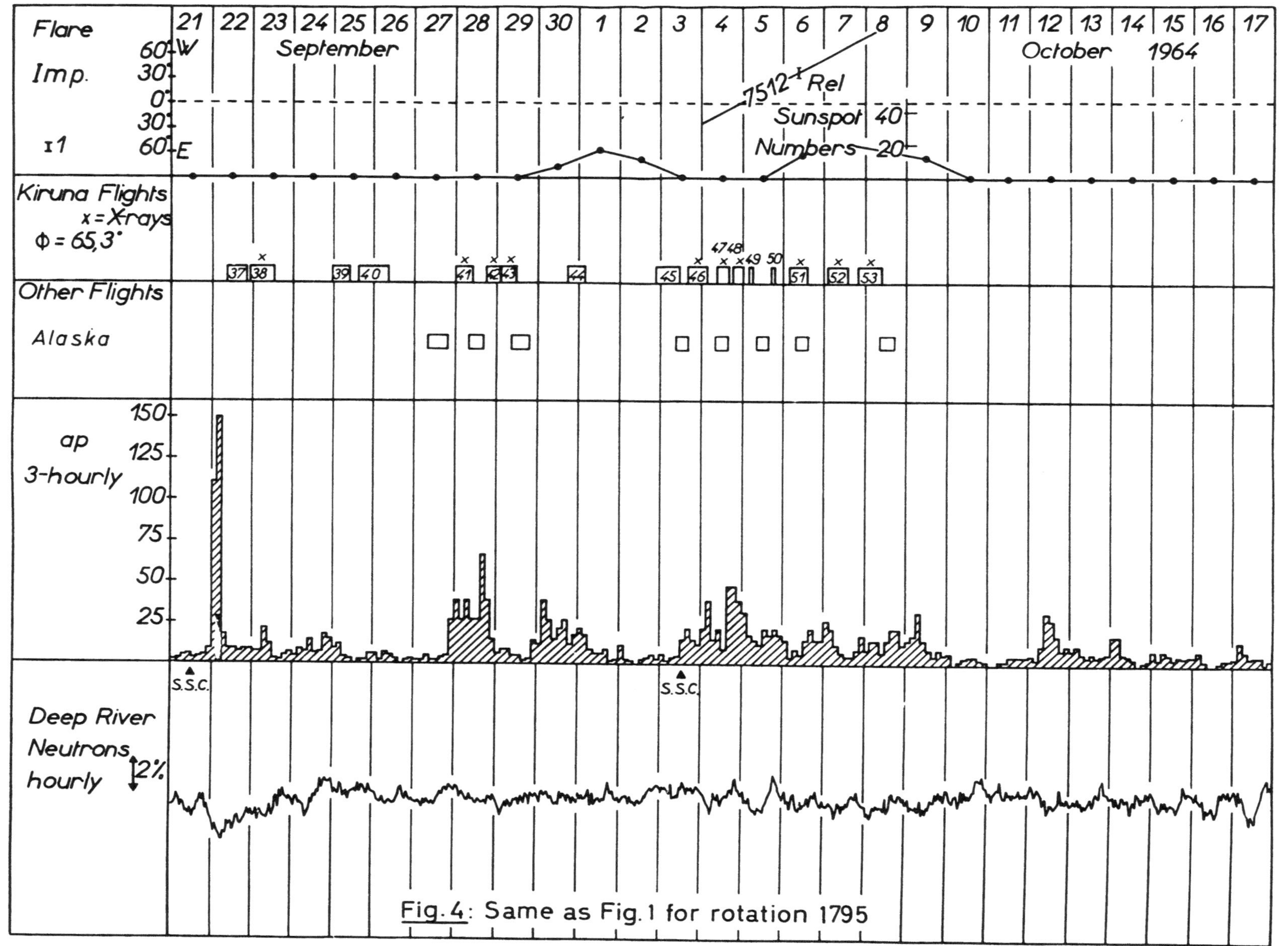

Fig. 4: Same as Fig. 1 for rotation 1795

Solar rotation 1794
(August 25 to September 20, 1964) (Fig. 3)

During this rotation four weak geomagnetically disturbed periods can be distinguished. The third of them was a geomagnetic storm beginning with an ssc on September 6 at 19.55 UT.

This disturbance and the fourth one were accompanied by small Forbush effects.

Flights K 21/64 to K 24/64 were launched during the second, flights K 27/64 to K 32/64 during the third, and flights K 33/64 to K 35/64 during the fourth disturbed period.

Solar rotation 1795
(September 21 to October 17, 1964 (Fig. 4)

The geomagnetic activity was higher than during the preceding rotations. During a storm beginning with an ssc on September 21 at 11.48 UT the ap's reached even a value of 179. This was also the only storm accompanied by a Forbush effect. The recurrent disturbed periods which developed from those of the preceding rotation lasted longer and partly overlapped. Flights K 37/64 to K 53/64 cover more or less closely this rotation until the end of the launching campaign.

Fig. 5 : Daily geomagnetic character figures C9 and sunspot number R. Symbolic tabulation in the 27-days sequence of the solar rotations after BARTELS. The recurrences of the M-regions are clearly seen. These diagrams are distributed by the Geophysikalisches Institut der Universität Göttingen, Germany, on request.

DAILY GEOMAGNETIC CHARACTER FIGURES C9 AND SUNSPOT NUMBERS R

Symbol	.	,	2	3	4	5	6	7	8	■
R =	0	1–15	16–30	31–45	46–60	61–80	81–100	101–130	131–170	171
C9 =	0	1	2	3	4	5	6	7	8	9
Cp =	0.0–0.1	0.2–0.3	0.4–0.5	0.6–0.7	0.8–0.9	1.0–1.1	1.2–1.4	1.5–1.8	1.9	2.0–2.5
Ap =	0–4	5–7	8–10	11–13	14–17	18–24	25–40	41–91	92–140	141–400

preliminary

IV. List of balloon launchings, Kiruna 1964.

Flight Nr.	Type of Detector	Start date	Start time (UT)	End of transmission date	End of transmission time (UT)	Ceiling pressure (mb)	balloon stayed above 50 mb pressure level from date	from time (UT)	to date	to time (UT)	Duration of flight above 50 mb pressure level (h).	Remarks
K 1/64	TESI	16.7	2200	17.7	1900	8	16.7	2320	17.7	1900	20	
K 2/64	TESCI	17.7	2020	18.7	0700	7	17.7	2142	18.7	0450	7	
K 3/64	TESCI	18.7	1938	18.7	2258	7.5	18.7	2115	18.7	2300	2	
K 4/64	TESI	19.7	2208	20.7	1300	6	19.7	2326	20.7	1300	14	
K 5/64	TESI	21.7	2110	22.7	0905	7	21.7	2224	22.7	0900	11	
K 6/64	TESCI	27.7	2120	28.7	0036	6	27.7	2243	27.7	2345	1	
K 7/64	TESCI	29.7	2110								-	failure
K 8/64	TESI	30.7	2112	31.7	1200	6	30.7	2226	31.7	1200	14	
K 9/64	TESI	2.8	2104	3.8	2030	6	2.8	2218	3.8	2030	22	
K 10/64	TESCI	3.8	2108								-	failure
K 11/64	TESI	4.8	2101	5.8	2032	6	4.8	2225	5.8	2030	22	
K 12/64	TESCI	5.8	2115	6.8	2000	9	5.8	2335	6.8	2000	21	
K 13/64	TESI	11.8	2005	12.8	2045	7	11.8	2135	12.8	2045	23	
K 14/64	TESCI	12.8	2002	13.8	1020	7	12.8	2115	13.8	1920	22	
K 15/64	TESCI	13.8	2000	14.8	2300	6	13.8	2115	14.8	2230	25	
K 16/64	TESCI	14.8	2145	15.8	0335	8	14.8	2306	15.8	0100	2	
K 17/64	TESI	18.8	2007	19.8	2117	6.5	18.8	2131	19.8	2115	24	
K 18/64	TESCI	20.8	2210								-	failure
K 19/64	TESI	25.8	1942	26.8	2020	8	25.8	2140	26.8	2020	22	
K 20/64	TESI	29.8	1912								-	failure
K 21/64	TESCI	30.8	1912	31.8	2200	7	30.8	2145	31.8	1925	22	
K 22/64	TESI	31.8	1910								-	failure
K 23/64	TESCI	31.8	2202	1.9	2020	7.5	31.8	2325	1.9	2000	21	
K 24/64	TESI	1.9	1948	2.9	2200	7	1.9	2109	2.9	2010	23	
K 25/64	TESI	3.9	2033								-	failure
K 26/64	TESI	4.9	1728	4.9	2000	8	4.9	1850	4.9	1947	1	
K 27/64	TESCI	5.9	1709	6.9	2310	5	5.9	1820	6.9	1900	25	
K 28/64	TESI	6.9	1810	6.9	2220	8	6.9	1945	6.9	2130	2	
K 29/64	TESI	6.9	2317	7.9	2030	10	7.9	0051	7.9	1900	18	
K 30/64	TESCI	7.9	1645	7.9	1922	14	7.9	1800	7.9	1850	1	
K 31/64	TESI	7.9	2051	8.9	1856	7	7.9	2210	8.9	1815	20	
K 32/64	TESI + I.C.	8.9	1843	9.9	1955	7	8.9	2010	9.9	1845	23	did not work after ceiling
K 33/64	TESCI	14.9	1855	15.9	1630	5	14.9	2005	15.9	?	?	
K 34/64	TESCI	15.9	2312	16.9	1800	7	16.9	0145	16.9	1740	16	
K 35/64	TESI	16.9	1918	17.9	0042	9	16.9	2039	16.9	2355	3	
K 36/64	TESCI	18.9	2212	19.9	0605	5	18.9	2329	19.9	0145	3	
K 37/64	TESI + I.C.	22.9	0947	22.9	2050	9	22.9	1103	22.9	1818	7	
K 38/64	TESIO + TESCI	22.9	2210	23.9	1325	6	22.9	2315	23.9	1325	14	
K 39/64	TESCI	25.9	0010	25.9	0850	6	25.9	0200	25.9	0850	7	
K 40/64	TESI	25.9	1536	25.9	2230	6	25.9	1645	25.9	2230	6	
K 41/64	TESCI	28.9	0019	28.9	1025	5	28.9	0135	28.9	1025	9	
K 42/64	TESI	28.9	1806	29.9	0130	7.5	28.9	1923	29.9	0130	6	
K 43/64	TESCI	28.9	2358	29.9	1055	6	29.9	0106	29.9	1055	10	
K 44/64	TESI	30.9	1805	1.10	0320	7	30.9	1911	30.9	2340	5	
K 45/64	TESI	2.10	2039	3.10	1000							no pressure measurement
K 46/64	TESCI	3.10	1551	4.10	0200	6	3.10	1700	4.10	0200	9	
K 47/64	TESI TESCI + I.C.	4.10	0954	4.10	1630	7	4.10	1054	4.10	1455	4	
K 48/64	TESI	4.10	1758	5.10	0030	7	4.10	1903	5.10	0030	5	
K 49/64	TESCI	5.10	0433	5.10	0645	14.5	5.10	0545	5.10	?	?	squibbed at 14.5 mb
K 50/64	TESI	5.10	1714	5.10	1927	10	5.10	1825	5.10	1905	0.5	
K 51/64	TESIO + TESCI	6.10	0413	6.10	1435	10	6.10	0520	6.10	1300	8	
K 52/64	TESI + TESCI	7.10	0333	7.10	1440	5	7.10	0440	7.10	1240	8	
K 53/64	TESCI	7.10	1955	8.10	1030	6.5	7.10	2118	8.10	1030	14	

V. Representation of the measurements.

The counting rates of all flights during which X-rays were measured are plotted versus time. In addition the counting rates of the single Geiger-Müller-tubes and the telescopes as well as of those scintillation counters, which descended just after ceiling, are plotted versus pressure.

The intensity versus time diagrams.

a)   The counting rates of the various detectors or of the different channels of the scintillation counters are plotted versus time on a logarithmic scale as given on both sides of the diagrams. Under each curve a base line is drawn which indicates how the logarithmic scale must be adjusted for this curve. The following notifications hold :

Al-GM :	Al-walled Geiger-Müller-tube (Victoreen 1 B 85),
Bi-GM :	Bi-coated Geiger-Müller-tube (Victoreen 6306),
Te :	Telescope,
I.C. :	Ionization chamber.

Curves, which are labeled with energy thresholds represent the counting rates due to different energy losses in the crystals of the scintillation counters. Corrected values for the thresholds as determined retrospectively after an inflight calibration are inserted in Table 2. (See remarks on the preflight and inflight calibration in section d.)

b)   The air pressure P at the balloon altitude is indicated by full dots. It is plotted also on the same logarithmic scale as the counting rates. In order to indicate the appropriate decade one pressure level is marked on each diagram.

c)   CNA-curve. During these flights the riometer did not always work satisfactorily. Sometimes the CNA-curve is therefore omitted or only indicated by a dashed line (short dashes). On most of the diagrams a quiet day absorption curve is tentatively drawn also by a dashed line (long dashes).

d)   Remarks on the calibration of the thresholds for the energy losses in the scintillation counters.

Preflight calibration.

A nominal setting of the thresholds was provided at 20 keV, 40 keV, 100 keV, and 500 keV. But these are only inserted in the diagrams, if they deviated by less than 10% from final preflight calibration values. This calibration was based on the 662 keV line of ^{137}Cs. On the base of a linear relation the thresholds were calculated from the ratio of the pulse height due to this line and those at which the discriminators for the different channels responded.

Inflight calibration

A probably more reliable calibration can be achieved by means of an average curve of the pulse height distribution due to the cosmic radiation at its secondary maximum .

Assuming that allowance can be made for the deviation of the cosmic ray intensity from a certain arbitrary chosen reference level, for instance on account of a neutron monitor record [ERBE, 1959] and assuming furthermore that the variation of the energy distribution and of the particle composition is a second order effect in this concern, the counting rates at the outputs of the different energy channels must be a unique function of the corresponding thresholds. Hence from the counting rates of the maximum, normalized after the above viewpoint, the actual threshold for each channel can be checked or respectively redetermined inflight.

This calibration function has been obtained retrospectively on a statistical base by plotting in Fig. 6 the counting rates of the maximum for times when no excess radiation was present, versus the original preflight values (see above). We took then the best fit curve as true representation of the relation between the actual counting rate and the actual threshold of each individual channel and used it for the evaluation of the corrected thresholds from the measured counting rates. These thresholds are inserted in Table 2 .

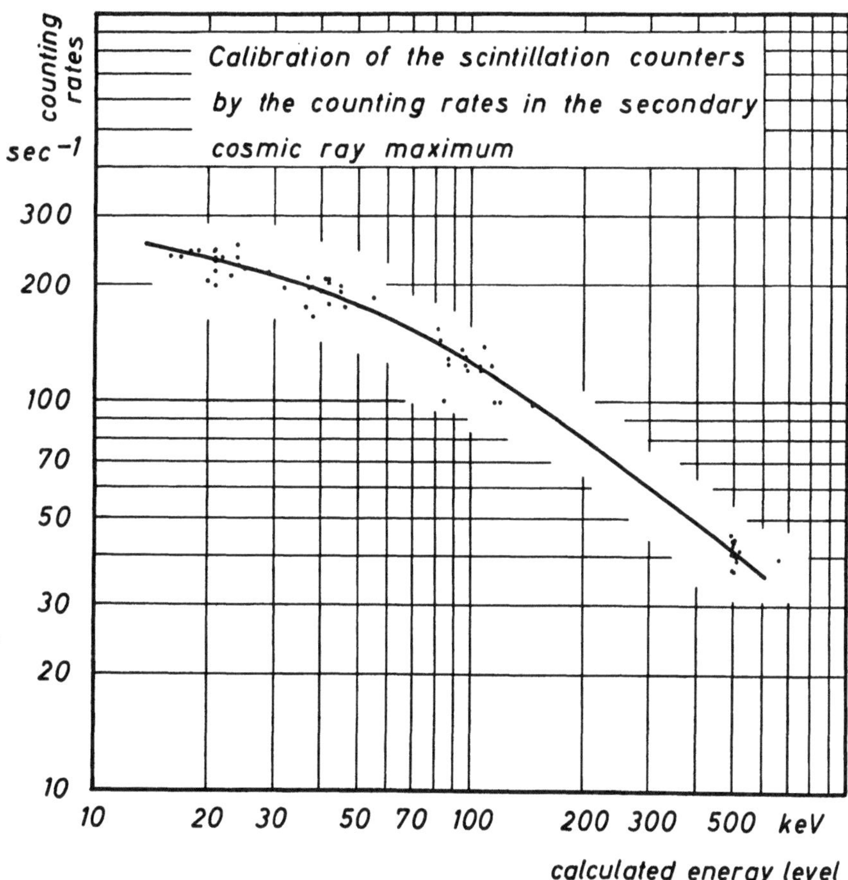

Fig. 6 :   The counting rates in the secondary cosmic ray maximum versus the calculated energy level. With the aid of this curve the energy levels (B) in Table 2 are deduced.

Table 2

Calibration of the scintillation counters by the counting rates in the secondary cosmic ray maximum

Flight Nr.	Nominal thresholds							
	20 keV		40 keV		100keV		500 keV	
	A	B	A	B	A	B	A	B
K 3/64	30	20	55	45	140	153	660	530
K 6/64	20	17	40	33	110	84	500	460
K 14/64	20	18	45	42	100	105	500	470
K 15/64	17	20	25	26	83	79	500	510
K 16/64	20	15	40	32	120	150	500	500
K 21/64	20	17	40	39	87	96	500	490
K 27/64	16	20	32	38	87	105	500	505
K 33/64	25	24	40	49	115	110	500	500
K 34/64	23	31	40	38	100	95	500	540
K 36/64	20	21	45	37	115	105	500	440
K 38/64	22	20	46	51	100	105	500	520
K 39/64	20	34	40	51	100	69	500	480
K 43/64	20	37	40	59	85	150	500	580
K 46/64	20	22	40	35	100	110	500	460
K 49/64	20	17	40	33	100	86	500	460
K 51/64	24	20	40	41	116	150	500	500

A = energy level marked on the diagrams (preflight calibration).

B = energy level after Fig. 6 (inflight calibration).

e)   The diagrams of the intensity versus pressure contain no additional curves. Here the intensity is plotted on a linear scale. The energy levels marked at the curves of the scintillation counters are those obtained by preflight calibration.

f)   Indications for launchings and comments concerning the flights.

K 1/64
Indication : Routine flight to test the equipment.
Two weak X-ray bursts were recorded.

K 2/64
Indication : Flights K 2/64 to K 4/64 were launched to cover an expected recurrence of an M-region (see comments on solar rotation 1792).

Great X-ray bursts were recorded when the balloon stayed above the 40 mb pressure level. All bursts were accompanied by AA-type CNA events.

K 3/64
Indication : see K 2/64.

One relatively soft X-ray burst occurred.

K 4/64
Indication : see K 2/64

The balloon was started during a disturbed period. Already at the secondary cosmic ray maximum additional radiation was measured, which continued until about 03 UT. Especially remarkable is the double-peaked event at about 01 UT, which is due to harder X-rays than the preceding and the following ones. It is accompanied by less enhanced absorption than the other events.

Another weak X-ray event occurred at about 10 UT.

K 5/64
Indication : Simultaneous SPARMO flight.

No additional radiation.

K 6/64
Indication : Routine flight.

Balloon was squibbed at 7 mb.

K 8/64
Indication : Routine flight.

Perhaps one weak X-ray event occurred just after 00 UT.

K 9/64
Indication : Flights K 9/64 to K 12/64 were launched to cover the period during which an M-region was expected to recur (see comments on solar rotation 1793).

One very small X-ray event occurred at about 23.40 UT. It was accompanied by a small peak in the CNA. This event occurred during the maximum (-80 γ) of a very smooth bay disturbance at Kiruna.

K 11/64
Indication : see K 9/64

Some very weak X-ray events were recorded. Remarkable is the single peak just after 20 UT on August 5. It was accompanied by a sudden increase of the CNA of 0,5 db and a sudden decrease of -215 γ , but only in the X-component of the geomagnetic field, which recovered slowly during about one hour. Though the sun was above the horizon at this time, we cannot ascribe this peak to solar γ-rays, as no special events were observed on the sun.

K 12/64
Indication : see K 9/64

X-rays were recorded during the total flight. They were all accompanied by enhanced CNA. But only during the first X-ray event a baylike geomagnetic disturbance was recorded at Kiruna. It is very remarkable that even during the great peak after 02 UT no distinct disturbance can be found on the magnetograms. According to KREMSERs findings (1964) it is very rare that a great X-ray event, which occurs only some hours after local midnight, is not accompanied by a baylike disturbance at the same station.

The events between 05 UT and 07 UT as well as around 10 UT are heavily structured. This structure can better be seen in the E > 40 keV channel than in the E > 20 keV channel, as the counting rates in the latter are scaled down by one more stage than in the E > 40 keV channel. This is also the reason why the peaks in the E > 40 keV channel are sometimes greater than in the E > 20 keV channel.

During the entire flight irregular geomagnetic pulsations with time intervals between 20 sec and 60 sec were recorded at Kiruna [WILHELM, Institut für Stratosphärenphysik am Max-Planck-Institut für Aeronomie, Lindau/Harz, private communication] . The peaks are sometimes coincident with those in the counting rates for intervals of 5 minutes and longer.

During this flight the counting rates of one of the Geiger-Müller-tubes, constituing the telescope, were transmitted to the ground. Thus, one can compare the counting rates of an Al-walled Geiger-Müller-counter with those of the scintillation counter. As expected, the less conspicuous events are not indicated by the Geiger-Müller-counter due to its relatively low efficiency for photons.

The long lasting enhanced absorption after the burst beginning at about 05 UT resembles a special type of events described by PFOTZER et al. (1965), and by KREMSER and BEWERSDORFF(1964) (see also K 21/64).

K 13/64
Indication : Flights K 13/64 to K 16/64 were started to cover the expected recurrence of an M-region (see comments on solar rotation 1793).

The abrupt increase of an X-ray burst at about 22.30 UT coincides with a rapid enhancement of the CNA (3 db) and an as rapid starting phase of a negative geomagnetic bay disturbance (-400 γ).

K 14/64
Indication : see K 13/64

During the flight X-rays were measured nearly continuously though this period was geomagnetically rather quiet.

K 15/64
Indication : see K 13/64

Weak X-ray events were recorded at the beginning and the end of the flight. It is very probable that variations of the counting rates in the E > 83 keV and E > 500 keV channels between these events are due to instrumental effects (see also K 23/64).

K 16/64

Indication : see K 13/64

Balloon descended quickly after having reached the 7 mb level.

K 17/64

Indication : Routine flight.

No additional radiation.

K 19/64

Indication : After some very quiet days a positive bay occurred on August 25 in the afternoon.

Only one small X-ray event was recorded.

K 21/64

Indication : Flights K 21/64 to K 24/64 were launched to cover the expected recurrence of an
M-region (see comments on solar rotation 1794).

Several X-ray events were measured. During the three smaller peaks between about 23 UT and
24 UT the CNA was not enhanced. A remarkably persisting enhancement is observed after the
event at about 02 UT. This points probably to its relationship with a certain group of events described by PFOTZER et al. (1965) (see also K 12/64).

Of all events only that between 12 UT and 14 UT was accompanied by a (positive) bay disturbance
at Kiruna.

K 23/64

Indication : see K 21/64

During the total flight X-rays were observed. The greatest peak at about 03 UT coincided with an
AA-type absorption and a medium geomagnetic bay disturbance (- 190 γ). The other events occurred during minor or no geomagnetic disturbances apart from the peak at about 17 UT. This
peak is perhaps a part of a stronger event, which began during a positive bay disturbance at Kiruna
(+200 γ). Furthermore, the highly structured burst between 08 UT and 09 UT is remarkable. The
counting rates in the E > 100 keV channel show sometimes variations a little bit more than due to
statistics, the origin of which is probably instrumental.

K 24/64

Indication : see K 21/64

Only one X-ray burst was registered, which was accompanied by an AA-type CNA event and a bay
disturbance (- 400 γ).

K 26/64

Indication : Magnetic disturbances.

Balloon was squibbed at 8 mb.

K 27/64

Indication : Flights K 27/64 to K 32/64 were launched to cover an expected recurrence of an M-
region (see comments on solar rotation 1794).

Though the period of this flight was magnetically very quiet, some X-ray events are measured.
During the ssc at 19.55 UT no additional radiation could be recorded as the balloon had already
descended deeper than the 50 mb pressure level.

K 28/64

Indication : see K 27/64

Though flights K 27/64 and K 28/64 were partly overlapping, additional radiation during the ssc could not be registered as the balloon was not yet high enough. The balloon was squibbed at 8 mb, so only a part of an X-ray event is observed.

K 29/64
Indication : see K 27/64

Until about 17 UT only weak X-rays were measured (the sensitivity of the riometer has changed at about 13.40 UT).

After 17 UT a greater X-ray event began, but the balloon descended and only a part of the event is registered. As can be seen on the CNA-curve a very intense and quickly rising event occurred at about 18.30 UT. This event is recorded by the balloon instruments of K 30/64, which was started when K 29/64 began to descend (see K 30/64).

K 30/64
Indication : see K 27/64, continuation of the events recorded at the end of K 29/64.

As one can see on the riometer curve at the end of K 29/64 a very intense electron precipitation event had occurred at about 18.30 UT. Though K 30/64 reached only the 10 mb level and then descended very quickly, the radiation during the most intense peaks of this event could be registered. The counting rates in all channels (only the $E > 18$ keV channel did not work) show a very quick increase (the time scale is expanded) exactly at the time of a CNA increase of about 5 db and a sudden decrease in the X-component of the geomagnetic field of about $- 500\gamma$. In the $E > 34$ keV channel the additional counting rate amounts to about 30 000/sec (625 photons/cm^2 sec sterad.) Even in the $E > 500$ keV channel additional radiation was measured.

When K 30/64 descended K 31/64 was launched, so it was possible to continue the radiation measurements with a lack of only 2 hours (see K 31/64).

K 31/64
Indication : see K 27/64, continuation of the events recorded at the end of K 29/64 and during K 30/64.

During the ascent the last part of the great precipitation event which began at the end of K 29/64 was registered. Then the geomagnetic activity diminished, but several important X-ray bursts were recorded until about 15 UT, when the balloon began to descend.

K 32/64
Indication : see K 27/64

The activity of this M-region diminished and only medium X-ray bursts could be observed. Remarkable is the event after 22 UT, which occurred during only very little enhancement of the CNA. Medium geomagnetic disturbances were recorded at that time at Kiruna.

K 33/64
Indication : Alert : "Magcalme geoalert".

No special event.

K 34/64
Indication : CNA and weak magnetic disturbances in the afternoon.

The greatest event occurred between 14 UT and 16 UT, when the X-ray activity at Kiruna is usually low.

K 35/64
Indication : Recurrence of geomagnetic disturbances was expected.

Balloon descended quickly after having reached the 9 mb level.

K 36/64

Indication : Magnetic micropulsations.

No special event.

K 37/64

Indication : Geomagnetic disturbances were expected.

No special event.

K 38/64

Indication : Alert : "Magstorm".

This flight shows well the differences between the scintillation counter and Geiger-Müller-counter results for events of medium intensity. During the bursts beginning at about 07.30 UT only the first one was also detected by the Geiger-Müller-counter. Small geomagnetic disturbances were recorded at Kiruna.

In this flight a scintillation counter and a TESIO were flown. (The TESIO worked only until the ceiling of the balloon). Due to this combination of instruments the counting rates of a vertically and a horizontally mounted Geiger-Müller-counter can be compared. The differences during the ascent are small. The counting rates of the vertically mounted Geiger-Müller-counter are lower than those of the horizontally mounted (compare K 51/64).

K 39/64

Indication : Micropulsation activity in the evening.

It stopped, however, soon and only very weak X-ray events occurred.

K 40/64

Indication : Events following a solar flare were expected.

No special event occurred.

K 41/64

Indication : Flights K 41/64 were launched during the expected recurrence of an M-region (see also comments on solar rotation 1795).

At the beginning of the flight a very intense X-ray burst occurred. During its decay phase great intensity pulsations were observed in the $E > 25$ keV and $E > 40$ keV channels which can also be found in the CNA curve.

All events occurred during a weak geomagnetic storm.

K 42/64

Indication : see K 41/64

Instruments worked unsatisfactorily, when the balloon had reached ceiling altitude.

K 43/64

Indication : see K 41/64.

After a geomagnetically disturbed afternoon and night, a quiet period began, when the balloon was launched. Nevertheless, several X-ray bursts could be observed.

K 44/64

Indication : see K 41/64

Balloon descended after ceiling. No special events could be observed.

K 45/64 (no figure).

Indication : The flights K 45/64 to K 53/64 were launched during the expected recurrence of an M-region (see also comments on solar rotation 1795).

Aurora was observed, however, no X-ray events. The intensity versus pressure curve could not be derived as the pressure unit did not work.

K 46/64
Indication : see K 45/64

The balloon descended very soon. Only weak X-ray events could be observed. There are no peaks in the counting rates at about 21.40 UT and 01.50 Ut, when the CNA suddenly increased. This is certainly due to the low altitude of the balloon already reached at these times.

K 47/64
Indication : see K 45/64

The last part of an X-ray event could be observed when the balloon reached its ceiling altitude. It descended, however, just afterwards.

K 48/64
Indication : see K 45/64

The balloon levelled during great geomagnetic disturbances. One X-ray event with great short-time intensity variations was observed. It was accompanied by strong additional CNA.

K 49/64
Indication : see K 45/64

Signals could only be received up to 16 mb. Until then no special events occurred.

K 50/64
Indication : see K 45/64

Balloon descended just after ceiling.

K 51/64
Indication : see K 45/64

One small and soft X-ray event occurred. For the combinations of instruments see K 38/64.

K 52/64
Indication : see K 45/64

One greater X-ray event was measured with a TESCI and a TESI. The differences in the amplitude and the time resolution can be well recognized.

The counting rates of the telescope of the TESI are lower than those of the TESCI. This is due to the heavier shielding of the telescope of the TESI, as this consists of two Al-walled and one Bi-coated Geiger-Müller-counters and that of the TESCI only of three Al-walled counters. Thus, a vertically incident particle must traverse 150 mg/cm^2 in the telescope of the TESCI, but 360 mg/cm^2 in the telescope of the TESI (see Table 1, p. 6).

K 53/64
Indication : see K 45/64

During the whole flight X-rays could be detected. At about 09 UT even the E > 440 keV channel showed additional radiation.

Summary

From July to October 1964, 53 balloon-flights were carried out at Kiruna to measure radiation associated with auroral zone disturbances. Different kinds of detectors were employed. Here the rough flight data are presented, which contain a great number of X-ray bursts. Some outstanding events are commented.

Acknowledgement

The French and German groups are deeply indebted to the ROYAL SWEDISH ACADEMY OF SCIENCE and its permanent secretary Prof. Dr. E. RUDBERG for permission to carry out the balloon launchings at the Kiruna Geophysical Observatory. They express their gratitude to the director of the observatory, Dr. B. HULTQVIST, and all colleagues and members of the observatory staff for their cordial and friendly cooperation.

All the authors appreciate the helpful efforts of the scientific and technical staff members of the cooperating institutions involved in the preparations of the instruments and the organisation of the campaign, and for the participation in the field operations.

We want to nominate especially the contributions of Dr. L. ROSSBERG by supervising the development of the TESCIs, as well as of Mr. G. SPITZ, by manufacturing and testing these instruments. The following technicians participated in the field operations:

Mr. Flentje, Mr. Fischer, Mr. Frank, Mr. Spitz (Lindau group)

Mr. Brendel, Mr. Jouan, Mr. Pagnier, Mr. Palous (Meudon group).

The research reported in this document has been sponsored in part:

through the Lindau group (Institut für Stratosphärenphysik am Max-Planck-Institut für Aeronomie), by the Bundesministerium für wissenschaftliche Forschung der Bundesrepublik Deutschland, by the Max-Planck-Gesellschaft granted mainly by the Deutsche Ländergemeinschaft and the Bundesrepublik Deutschland,

through the Kiruna group (Kiruna Geophysical Observatory, Director Dr. B. HULTQVIST), by the Air Force Cambridge Research Laboratories under Contract AF 61 (052)-604, through the European Office of Aerospace Research (OAR), United States Air Force,

through the Meudon group (Laboratoire de Physique Cosmique, Director Prof. Dr. A. FRÉON) by the Centre National d'Études Spatiales (CNES), France.

The factory Albin SPRENGER KG., St-Andreasberg, took over the manufacturing of the balloon equipments in series and we owe thanks to its representatives Dr. Chr. BACHEM and Mr. W. ULLRICH for their cordial response to all our requirements.

References

BARTELS, J. :   Terrestrial Magnetic Activity and its Relations to Solar Phenomena
Terr. Magn. Electr. $\underline{37}$, 1-52 (1932).

ERBE, H. :   Auswirkung der Variationen der primären kosmischen Strahlung auf die Mesonen-und
Nukleonenkomponente am Erdboden
Mitteilung a. d. Max-Planck-Institut für Aeronomie, Nr. 2, Springer-Verlag, Berlin,
Heidelberg, New-York, (1959)

KEPPLER, E. :   Description and Instruction Manual for the Sonde SPARMO 64
SPARMO Bulletin, Nr. 3, October 1964.

KREMSER, G. :   Über den Zusammenhang zwischen Röntgenstrahlungs-Ausbrüchen in der Polarlicht-
zone und bayartigen erdmagnetischen Störungen
Mitteilung a. d. Max-Planck-Institut für Aeronomie, Nr. 14, Springer-Verlag, Berlin,
Heidelberg, New-York (1964).

KREMSER, G. und BEWERSDORFF, A. :
Über eine besondere Art von Röntgenstrahlungs-Ausbrüchen in der Polarlichtzone
Kleinheubacher Berichte 1964.

PFOTZER, G. , EHMERT, A. , ERBE, H. , KEPPLER, E. , HULTQVIST, B. , and ORTNER, J. :
A Contribution to the Morphology of X-ray Bursts in the Auroral Zone
J. Geophys. Res. $\underline{67}$, 575-585, (1962a).

PFOTZER, G. , EHMERT, A. , and KEPPLER, E. :
Time Pattern of Ionizing Radiation in Balloon Altitudes in High Latitudes
Mitteilung a. d. Max-Planck-Institut für Aeronomie, Nr. 9, Springer-Verlag, Berlin,
Heidelberg, New-York (1962b).

PFOTZER, G. , EHMERT, A. , BEWERSDORFF, A. , KREMSER, G. , ROSSBERG, L. , RIEDLER, W. ,
TREFALL, H. , LEGRAND, J. P. :
Measurements of High Energetic Auroral Radiations with Balloon-Borne Detectors
in 1962 and 1963.
Mitteilung a. d. Max-Planck-Institut für Aeronomie, Nr. 18, Springer-Verlag, Berlin,
Heidelberg, New-York (1965).

ROSSBERG, L. and SPITZ, G. :
Eine Ballonsonde zur Messung des Energiespektrums von Röntgenstrahlungs-Schauern
und zum Nachweis geladener Teilchen (to be published).

<u>VI. Diagrams of the flights</u>

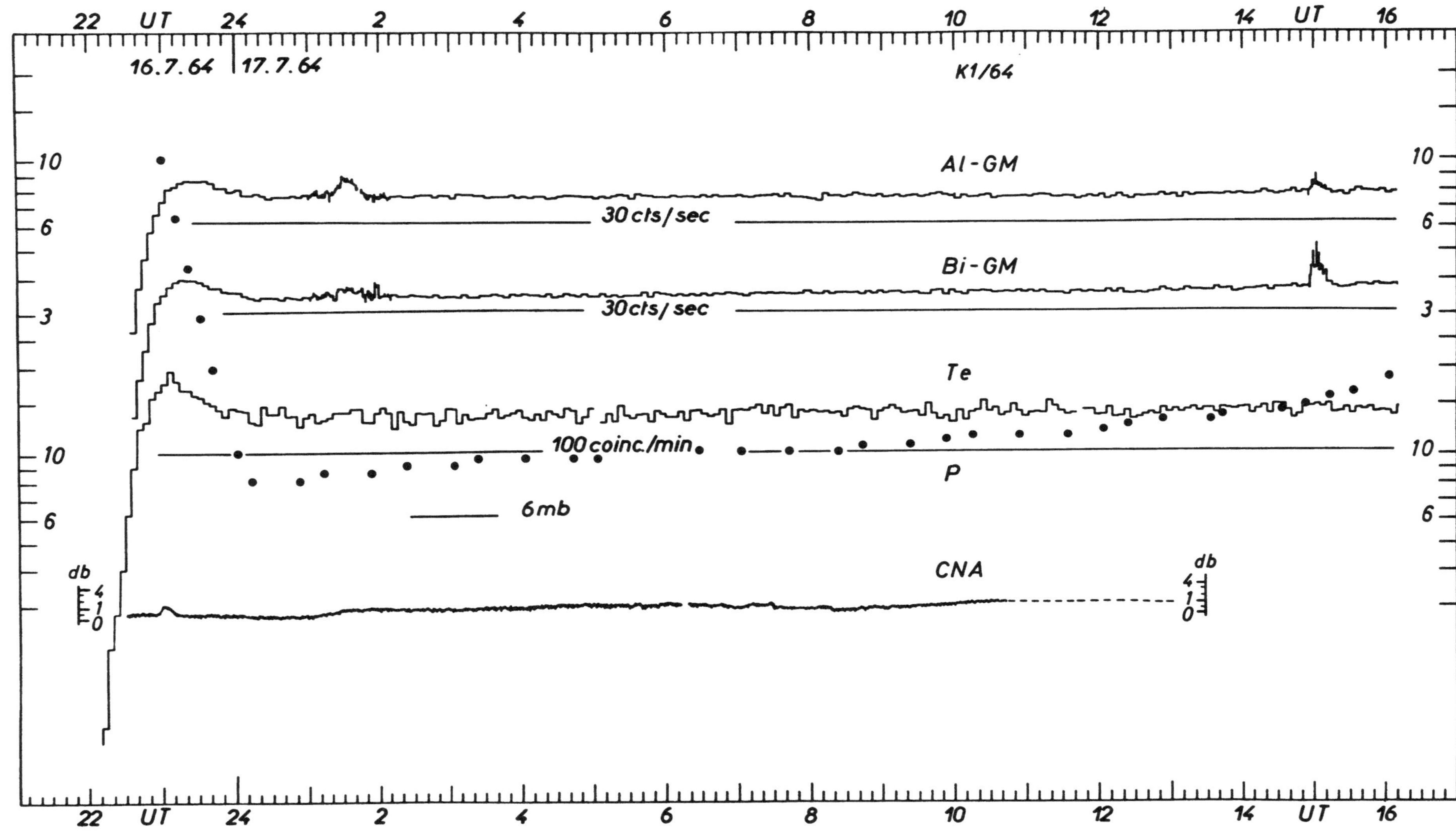
22 UT 24 2 4 6 8 10 12 14 UT 16
16.7.64 | 17.7.64
K1/64
Al-GM
30 cts/sec
Bi-GM
30 cts/sec
Te
100 coinc./min.
P
6mb
db
4
1
0
CNA
db
4
1
0
22 UT 24 2 4 6 8 10 12 14 UT 16

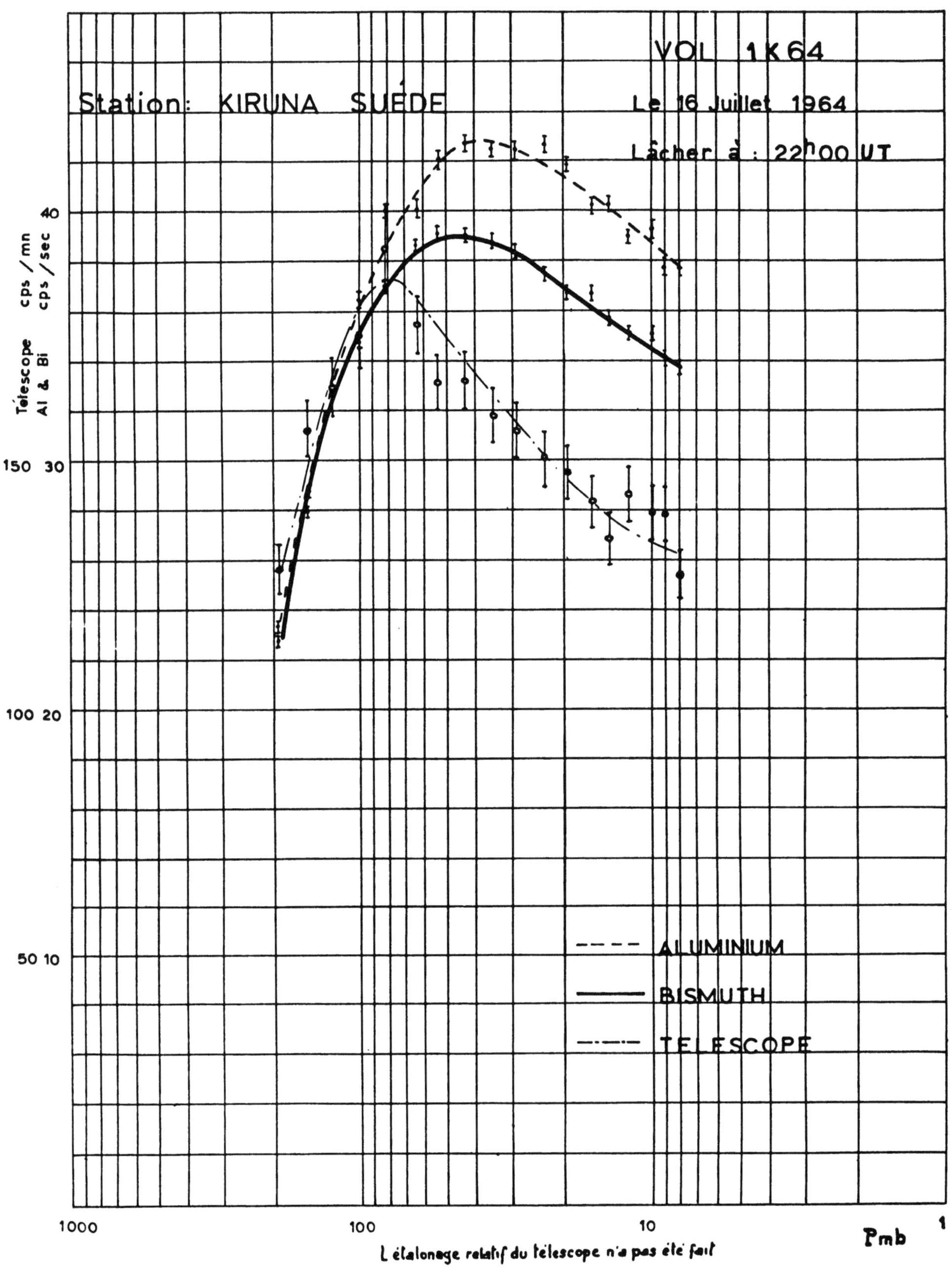

VOL 1 K 64
Station: KIRUNA SUÈDE
Le 16 Juillet 1964
Lâcher à : 22h00 UT
Téléscope cps / mn
Al & Bi cps / sec
40
150 30
100 20
50 10
----- ALUMINIUM
BISMUTH
-·-·- TELESCOPE
1000
100
10
Pmb
1
L'étalonage relatif du télescope n'a pas été fait

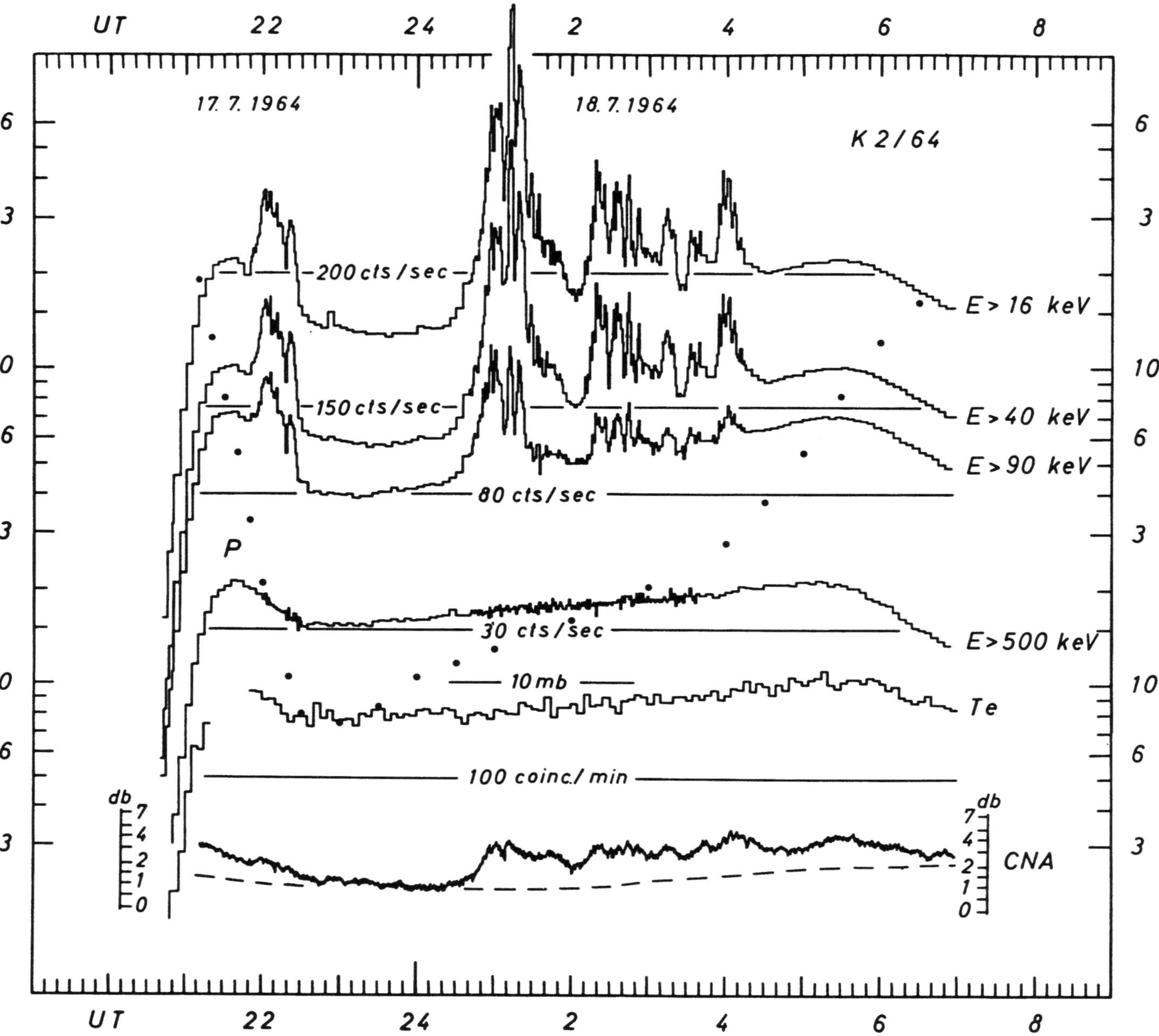

UT 22 24 2 4 6 8
17. 7. 1964
18. 7. 1964
K 2/64
200 cts/sec
E > 16 keV
150 cts/sec
E > 40 keV
E > 90 keV
80 cts/sec
P
30 cts/sec
E > 500 keV
10 mb
Te
100 coinc./min
db 7 4 2 1 0
db 7 4 2 1 0
CNA
UT 22 24 2 4 6 8

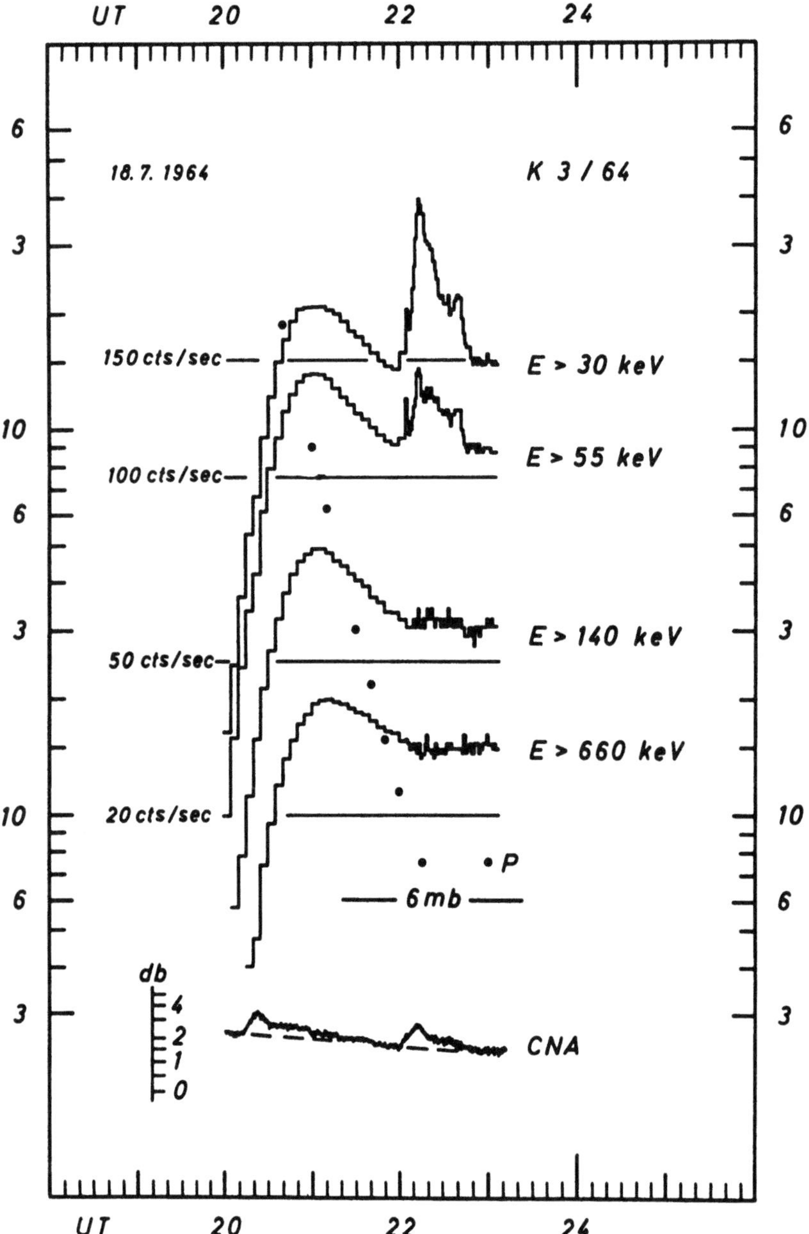

UT   20   22   24
18. 7. 1964
K 3 / 64
150 cts/sec
E > 30 keV
E > 55 keV
100 cts/sec
E > 140 keV
50 cts/sec
E > 660 keV
20 cts/sec
P
6 mb
db
4
2
1
0
CNA
UT   20   22   24

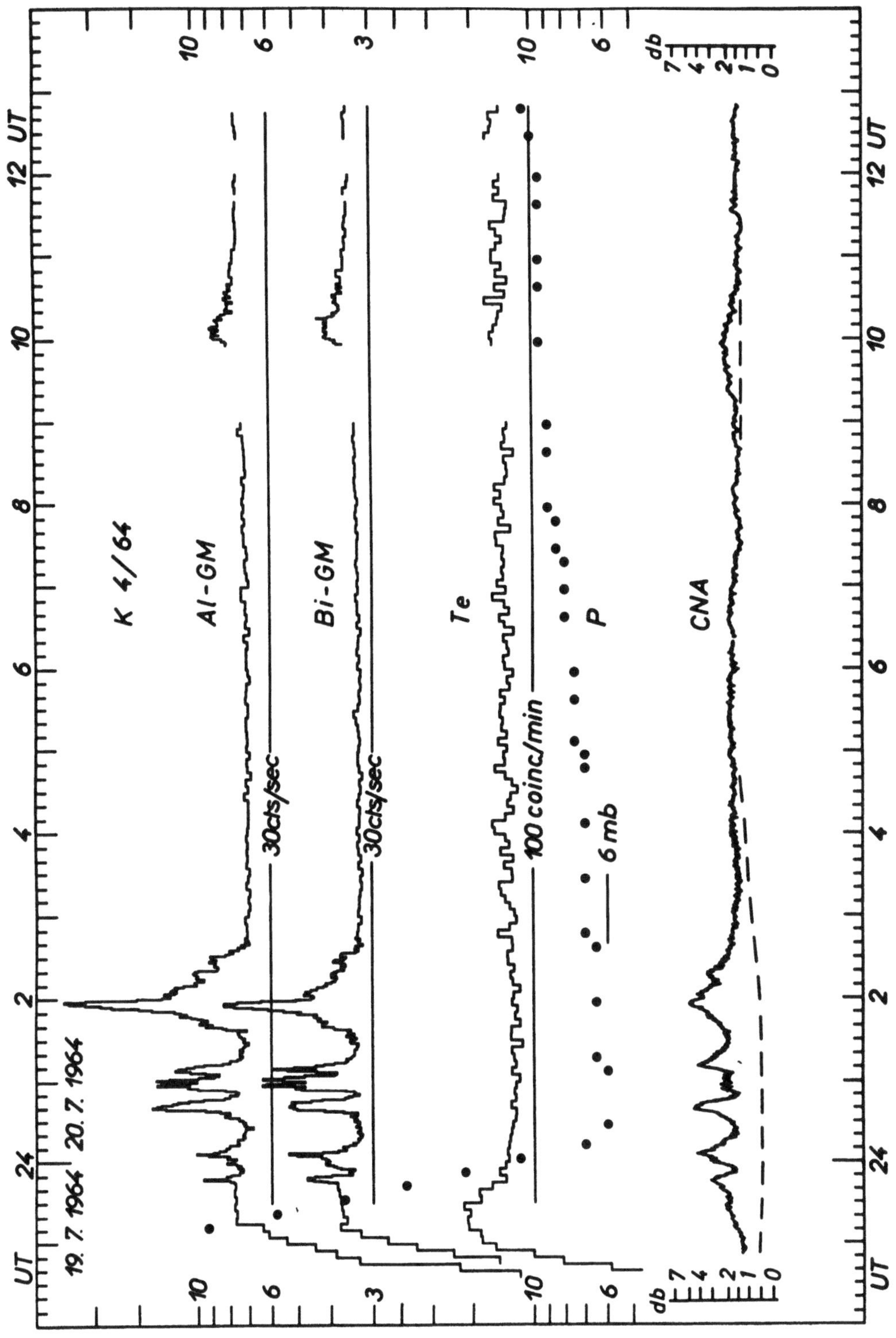

K 4/64
Al-GM
Bi-GM
Te
P
CNA
30 cts/sec
30 cts/sec
100 coinc/min
6 mb
UT
12 UT
19.7.1964
20.7.1964
10
6
3
10
6
db
7
4
2
1
0

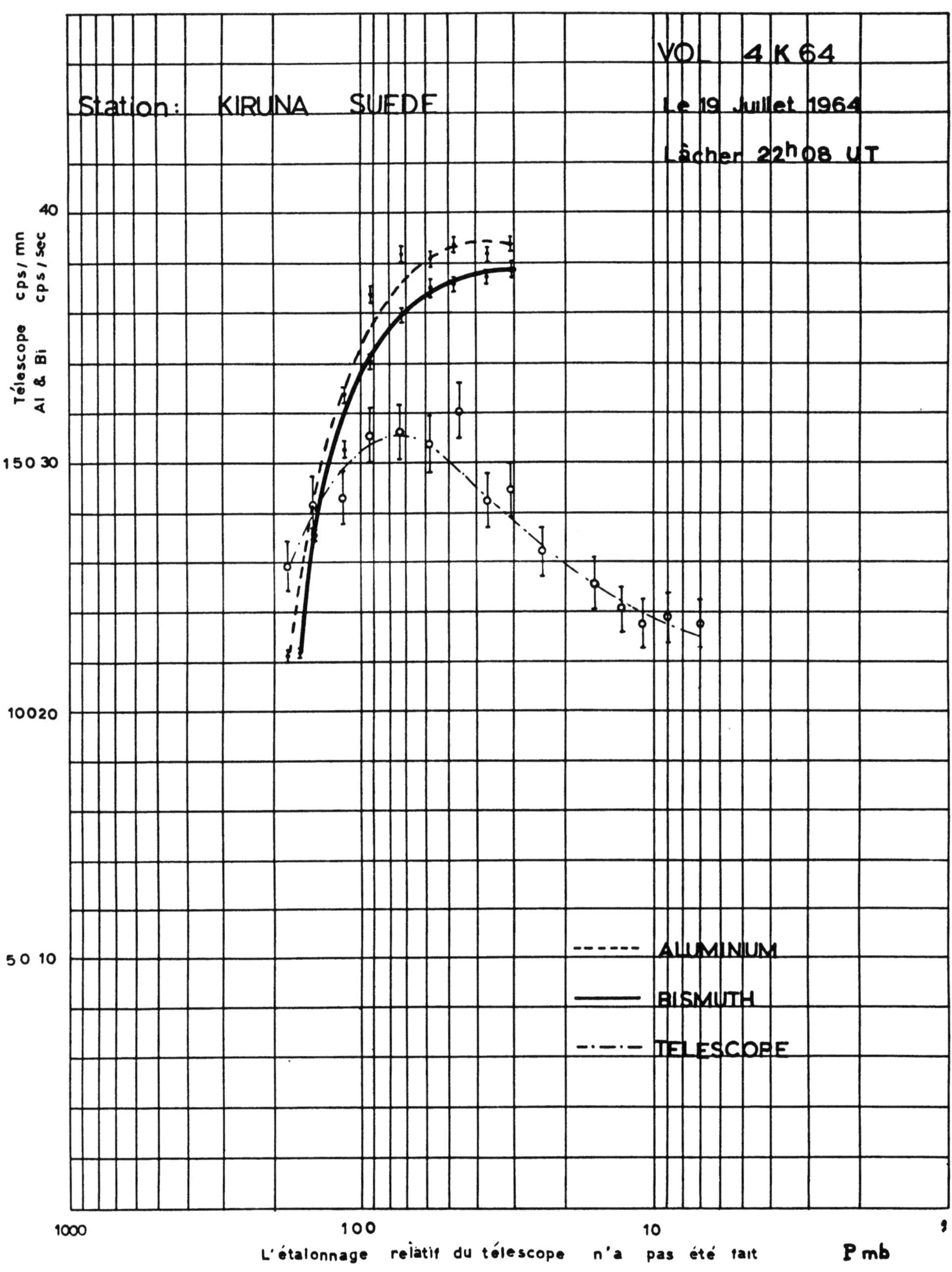

VOL 4 K 64
Station: KIRUNA SUEDE
Le 19 Juillet 1964
Lâcher 22h 08 UT
Télescope cps / mn
Al & Bi   cps / sec
40
30
150 30
100 20
50 10
ALUMINIUM
BISMUTH
TELESCOPE
1000
100
10
L'étalonnage relatif du télescope n'a pas été fait
P mb

K 5 / 64

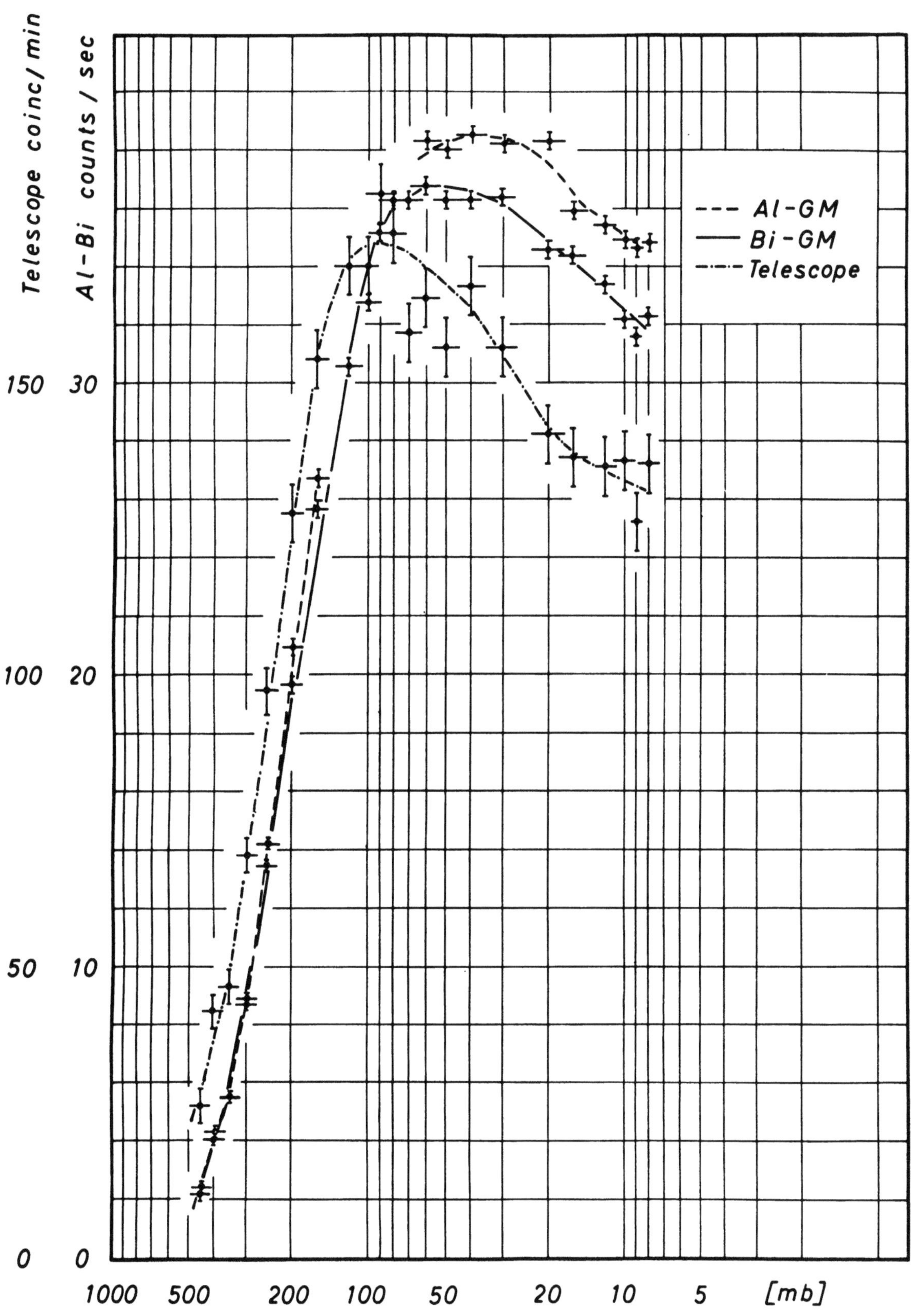

launched on July 21st 1964 at 21¹⁰ UT

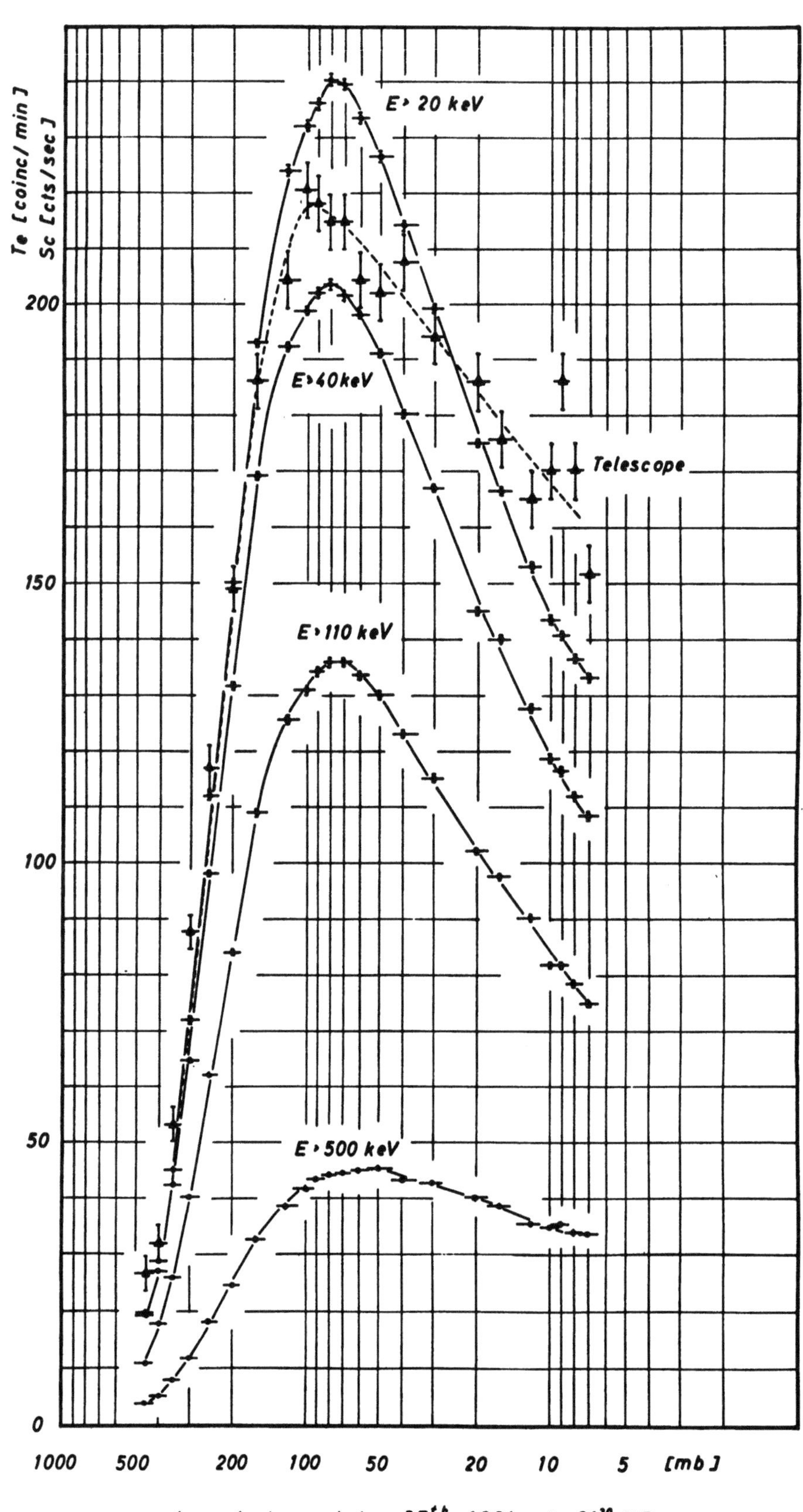

K 6 / 64
Te [coinc / min]
Sc [cts / sec]
200
150
100
50
0
E > 20 keV
E > 40 keV
E > 110 keV
E > 500 keV
Telescope
1000 500 200 100 50 20 10 5 [mb]
launched on July 27th 1964 at 21 20 UT

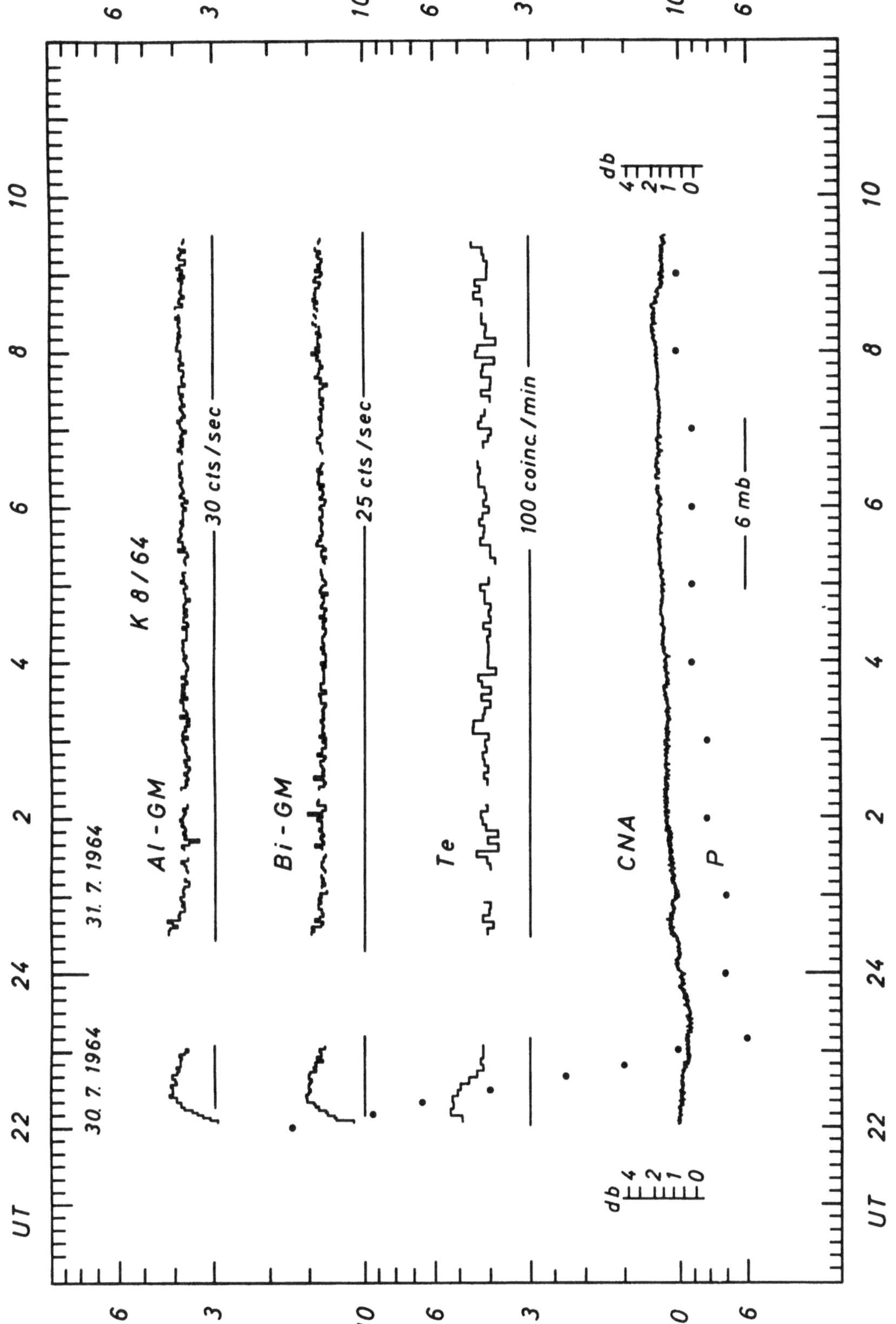

K 8/64
31. 7. 1964
30. 7. 1964
Al - GM
30 cts/sec
Bi - GM
25 cts/sec
Te
100 coinc./min
CNA
P
6 mb
db
4
2
1
0
UT
6
3
10
6
3
10
6

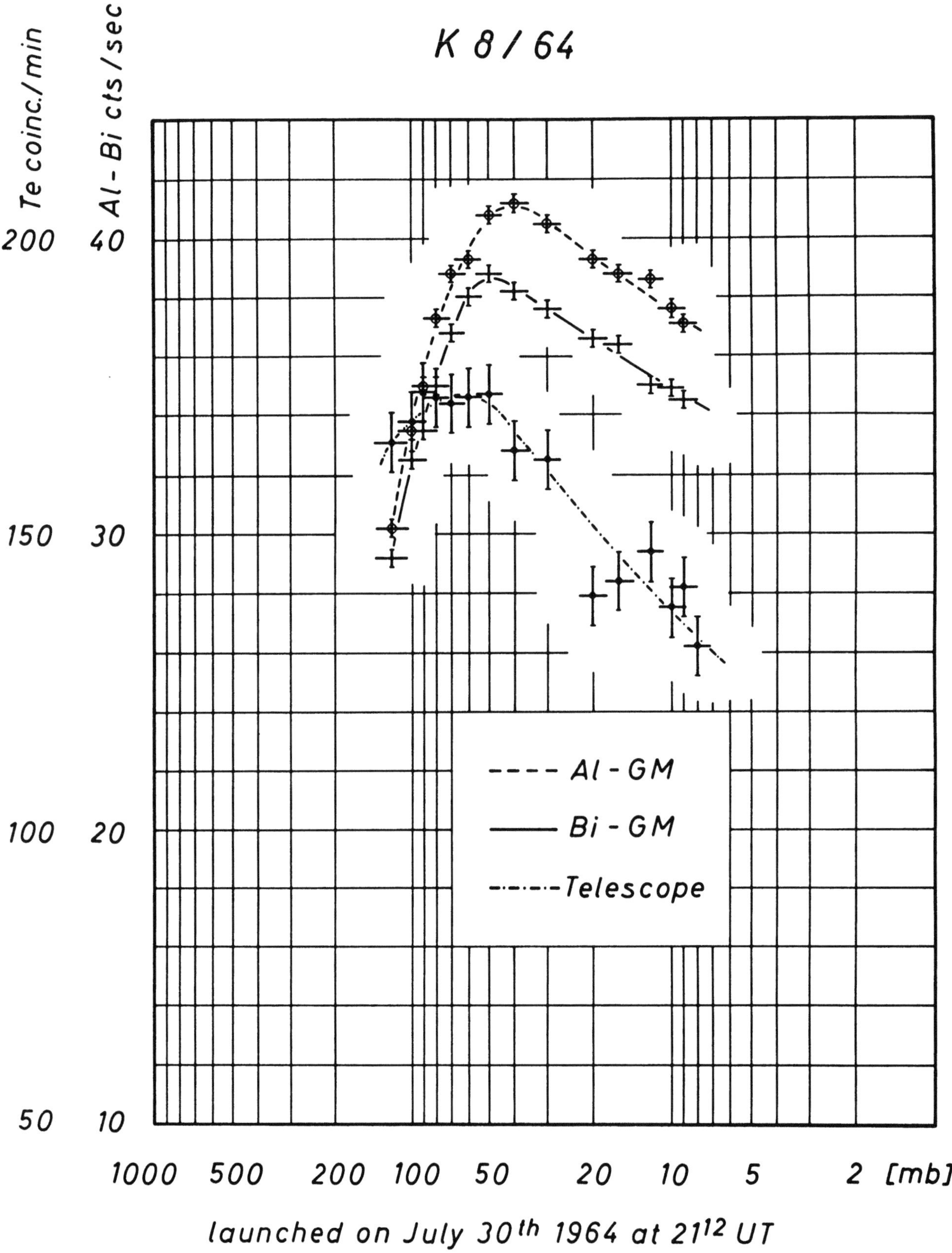

K 8 / 64
Te coinc./min
Al - Bi cts/sec
200
40
150
30
100
20
50
10
Al - GM
Bi - GM
Telescope
1000 500 200 100 50 20 10 5 2 [mb]
launched on July 30th 1964 at 21^{12} UT

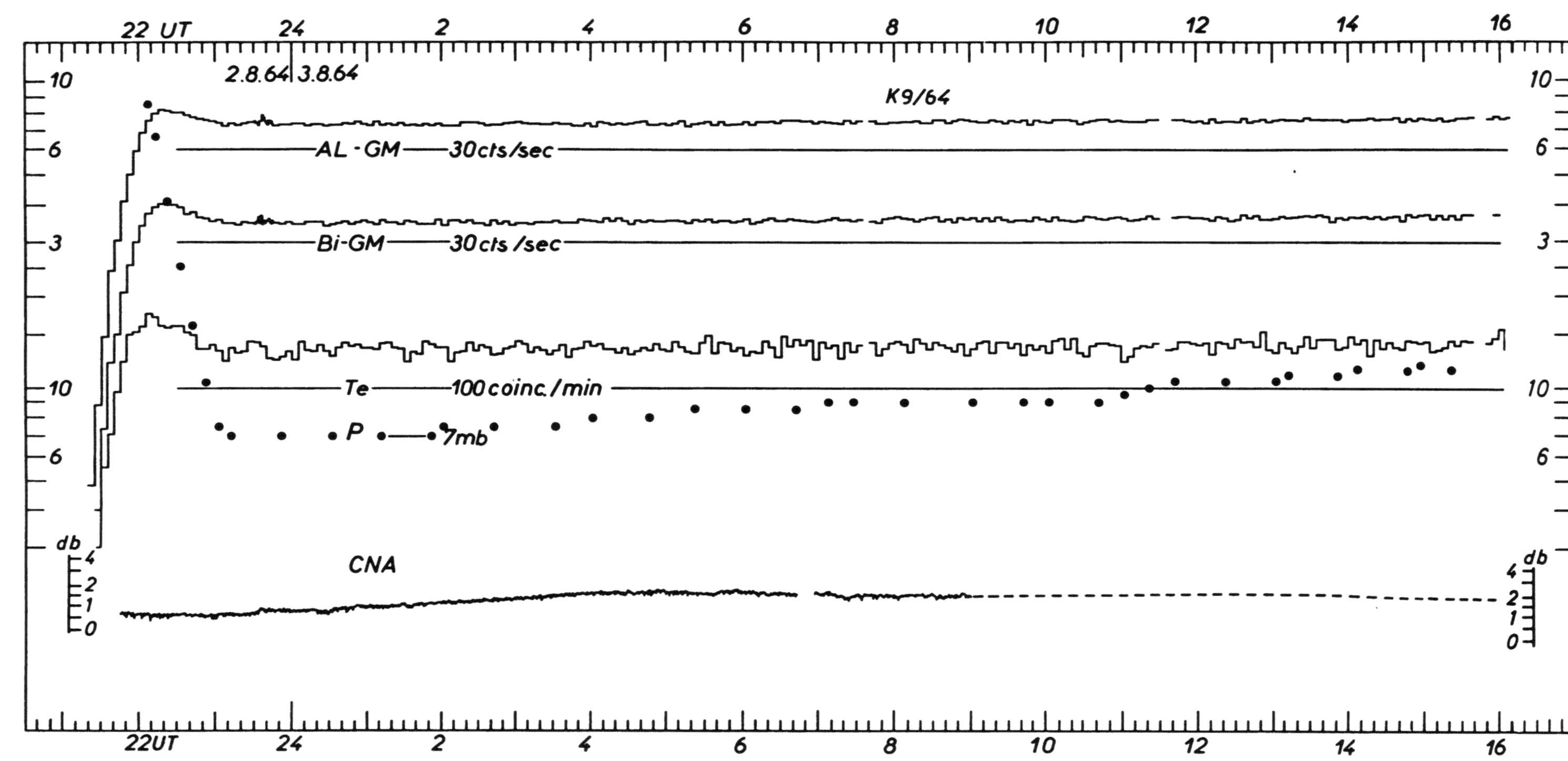

22 UT   24   2   4   6   8   10   12   14   16
10
2.8.64 3.8.64
K9/64
6
AL · GM ——30cts/sec
3
Bi-GM ——30cts /sec
Te ——100 coinc./min
10
P ●——7mb
6
db
4
2
1
0
CNA
22UT   24   2   4   6   8   10   12   14   16

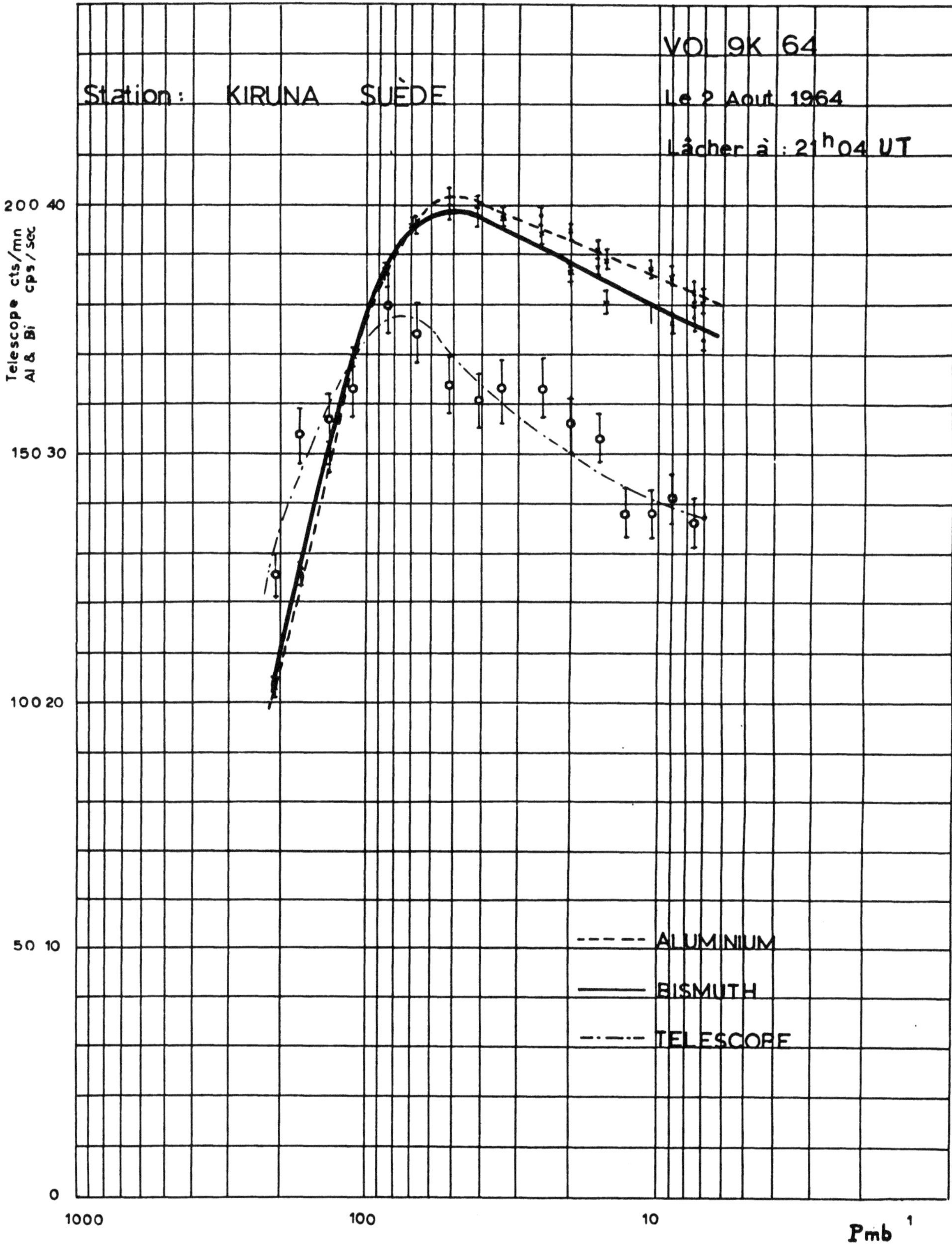

VOL 9K 64
Station : KIRUNA SUÈDE
Le 2 Aout 1964
Lâcher à : 21h04 UT
Telescope cts/mn
Al & Bi cps/sec
200 40
150 30
100 20
50 10
0
1000
100
10
Pmb
ALUMINIUM
BISMUTH
TELESCOPE

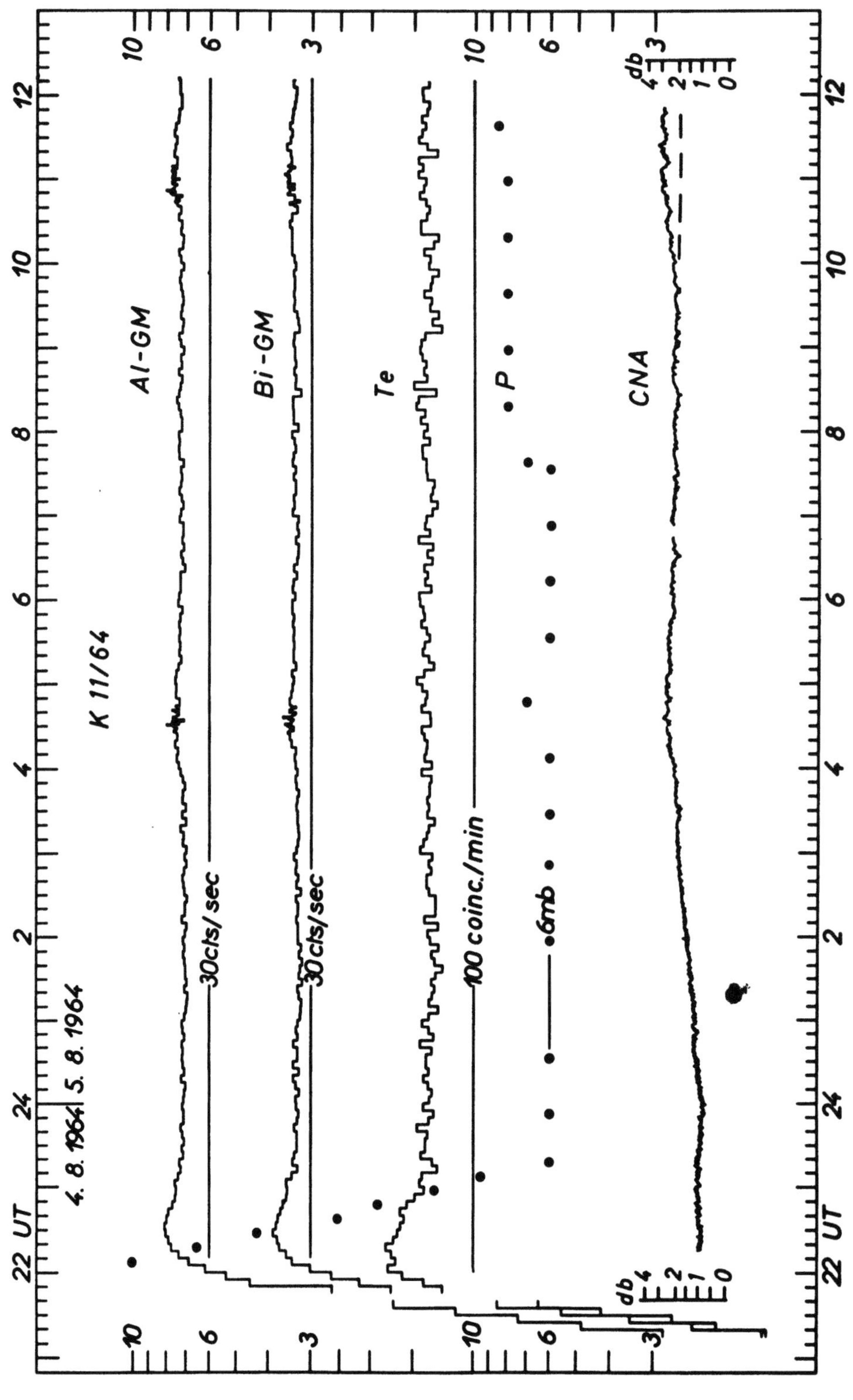

K 11/64
4. 8. 1964
5. 8. 1964
AI-GM
30 cts/sec
10
6
Bi-GM
30 cts/sec
3
Te
100 coinc./min
10
6
P
6 mb
CNA
db
4
3
2
1
0
22 UT
24
2
4
6
8
10
12

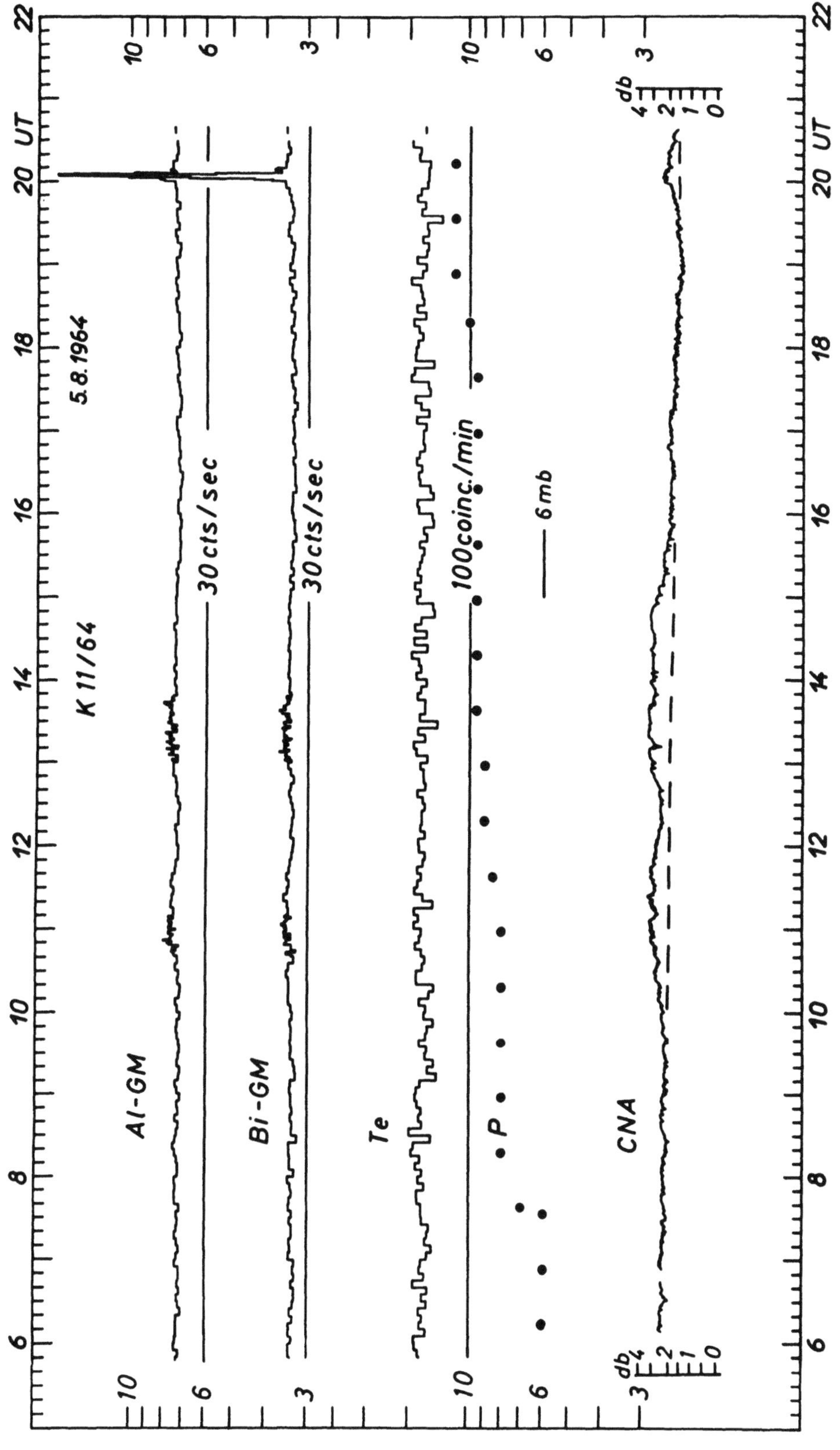

5.8.1964
K 11/64
Al-GM
30 cts/sec
Bi-GM
30 cts/sec
Te
P
100 coinc./min
6 mb
CNA
22 20 UT 18 16 14 12 10 8 6
10 6 3
10 6
db 4 3 2 1 0

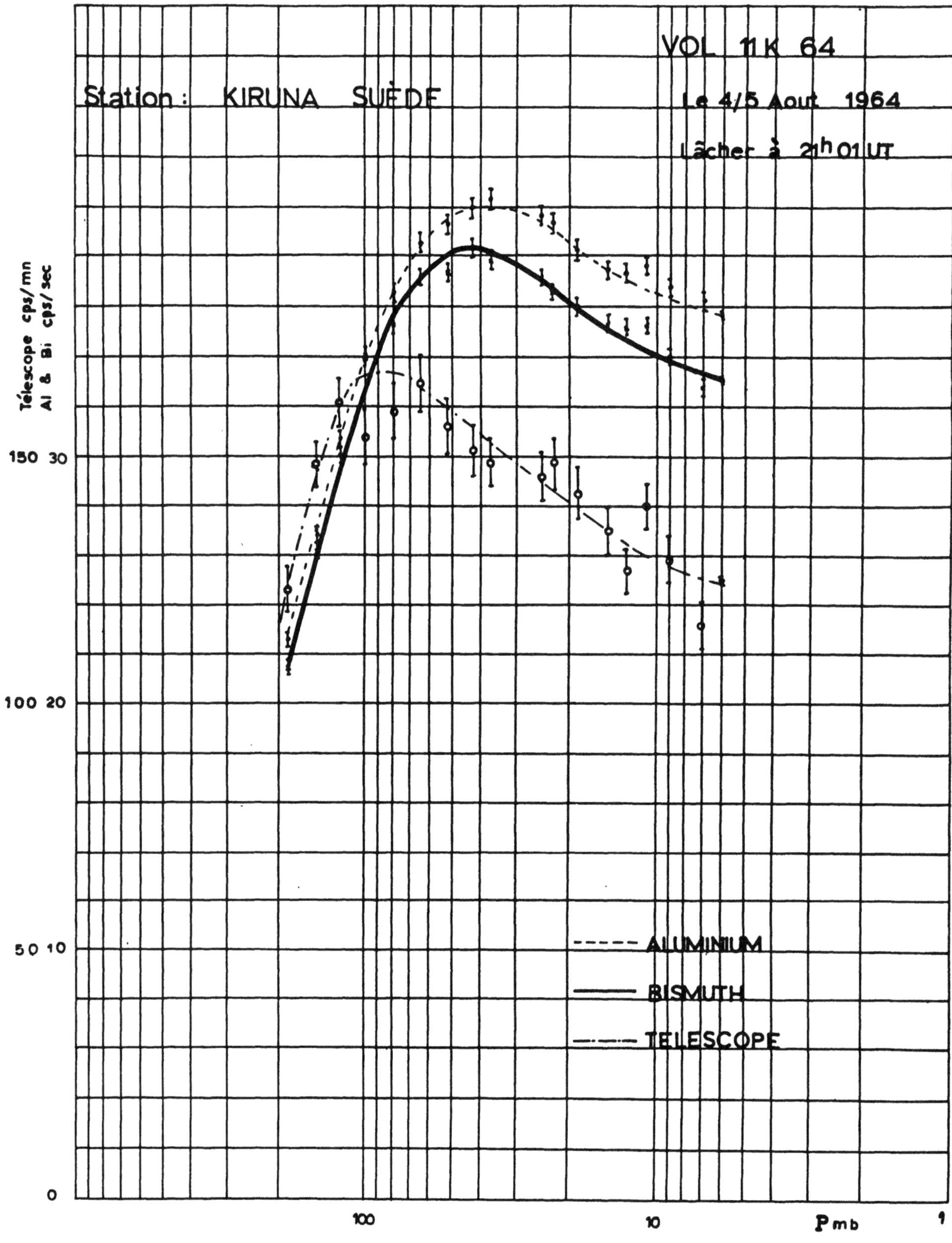

VOL 11 K 64
Station : KIRUNA SUÈDE
Le 4/5 Aout 1964
Lâcher à 21h 01 UT
Télescope cps/mn
Al & Bi cps/sec
150 30
100 20
50 10
0
ALUMINIUM
BISMUTH
TELESCOPE
100
10
P mb
1

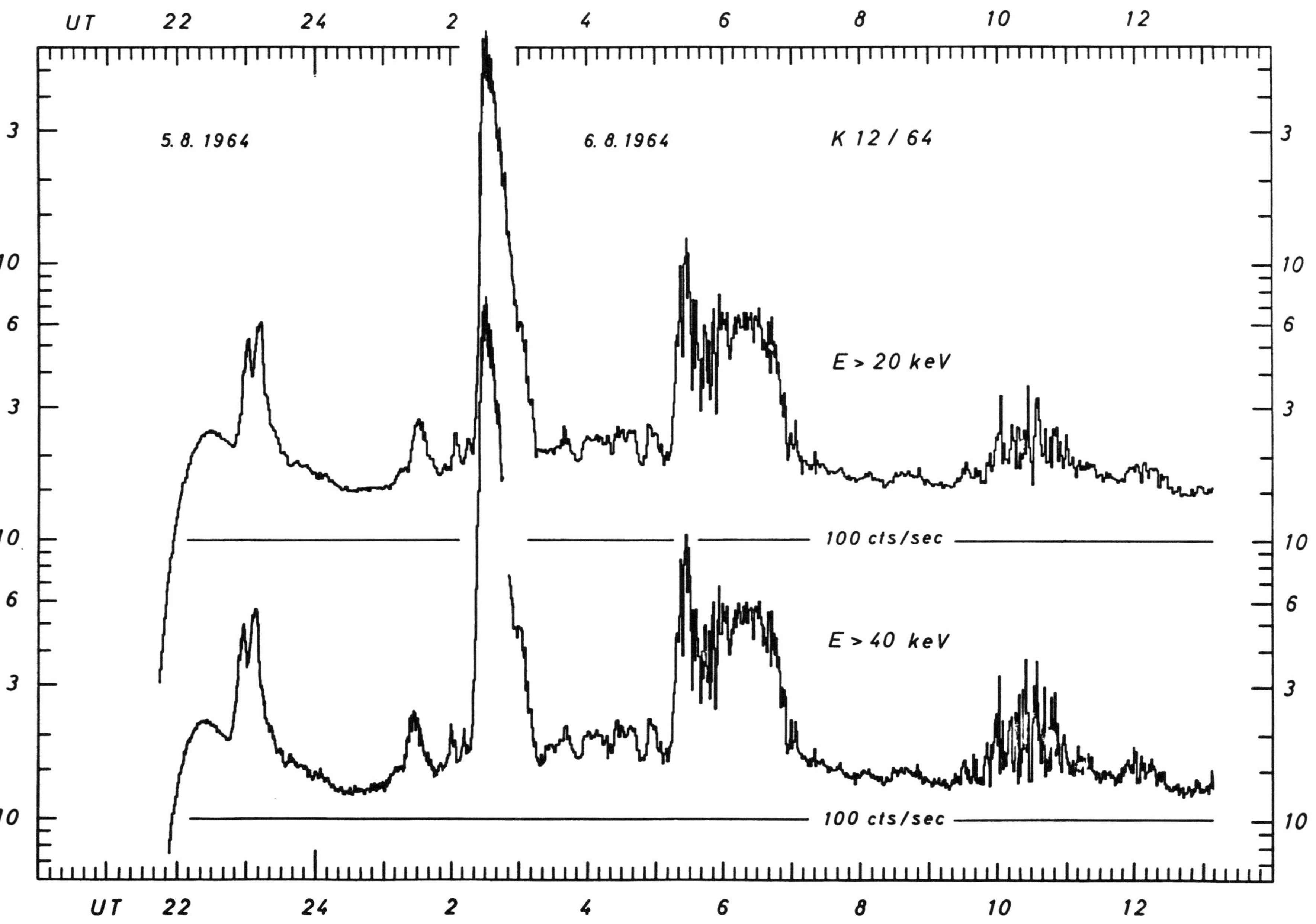

UT 22 24 2 4 6 8 10 12
5. 8. 1964
6. 8. 1964
K 12 / 64
E > 20 keV
100 cts/sec
E > 40 keV
100 cts/sec
UT 22 24 2 4 6 8 10 12
3 10 6 3 10 6 3 10

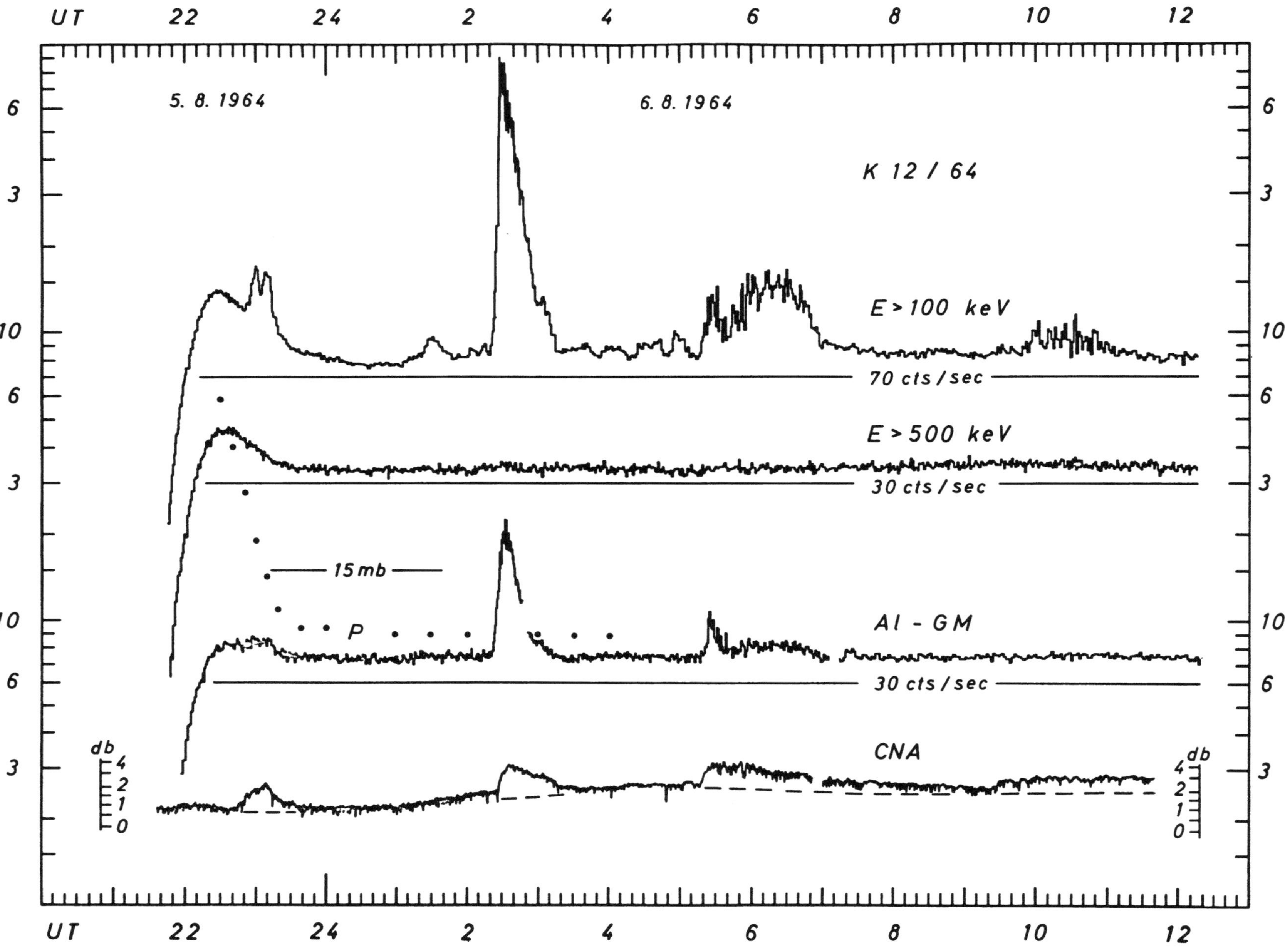

UT
22
24
2
4
6
8
10
12
5. 8. 1964
6. 8. 1964
K 12 / 64
E > 100 keV
70 cts / sec
E > 500 keV
30 cts / sec
15 mb
P
Al - GM
30 cts / sec
CNA
db
4
2
1
0
UT

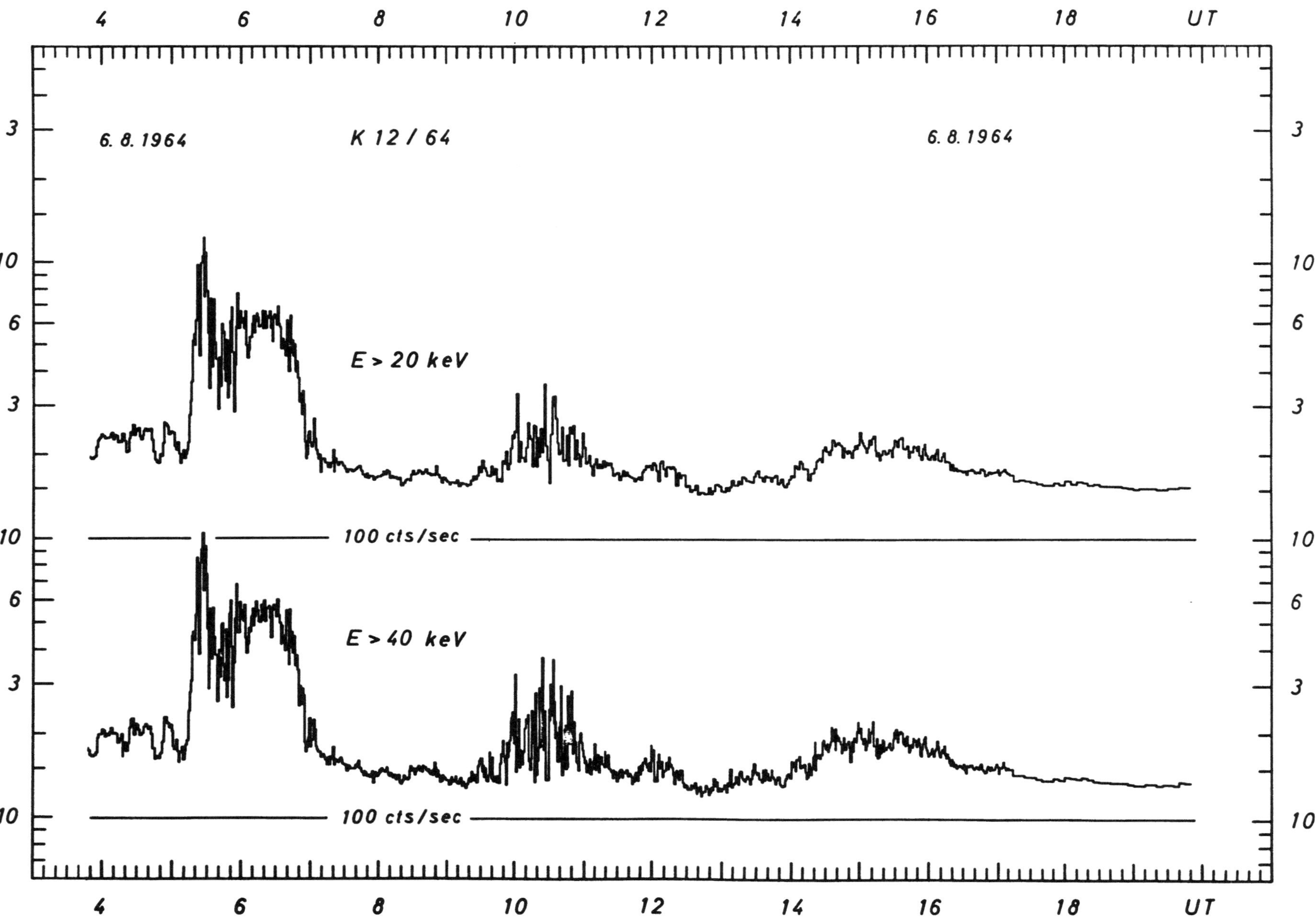

4 6 8 10 12 14 16 18 UT
6. 8. 1964
K 12 / 64
6. 8. 1964
E > 20 keV
100 cts/sec
E > 40 keV
100 cts/sec
3 10 6 3 10 6 10

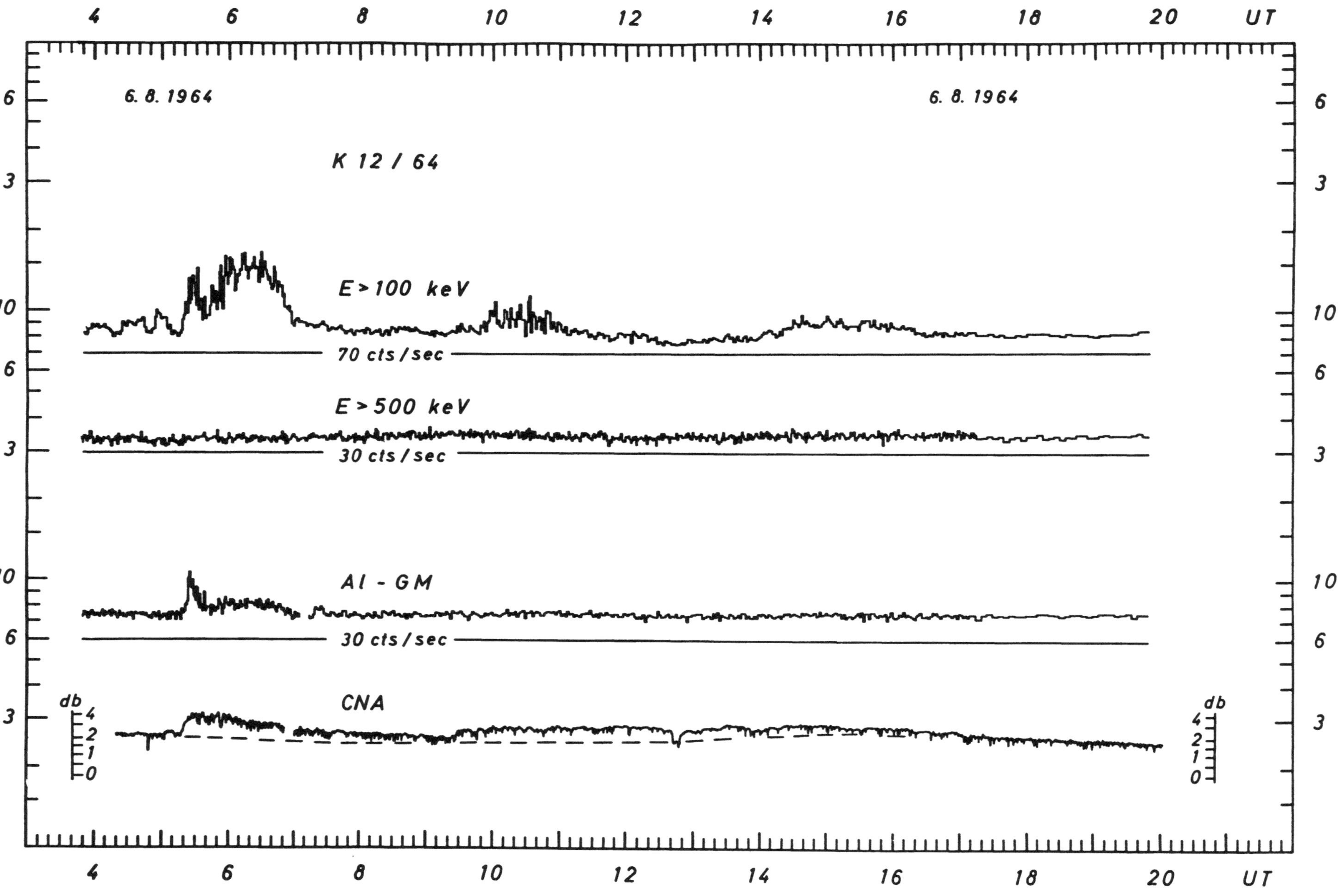

6. 8. 1964
6. 8. 1964
K 12 / 64
E > 100 keV
70 cts / sec
E > 500 keV
30 cts / sec
AI - GM
30 cts / sec
CNA
4 6 8 10 12 14 16 18 20 UT
4 6 8 10 12 14 16 18 20 UT
db 4 2 1 0
db 4 2 1 0

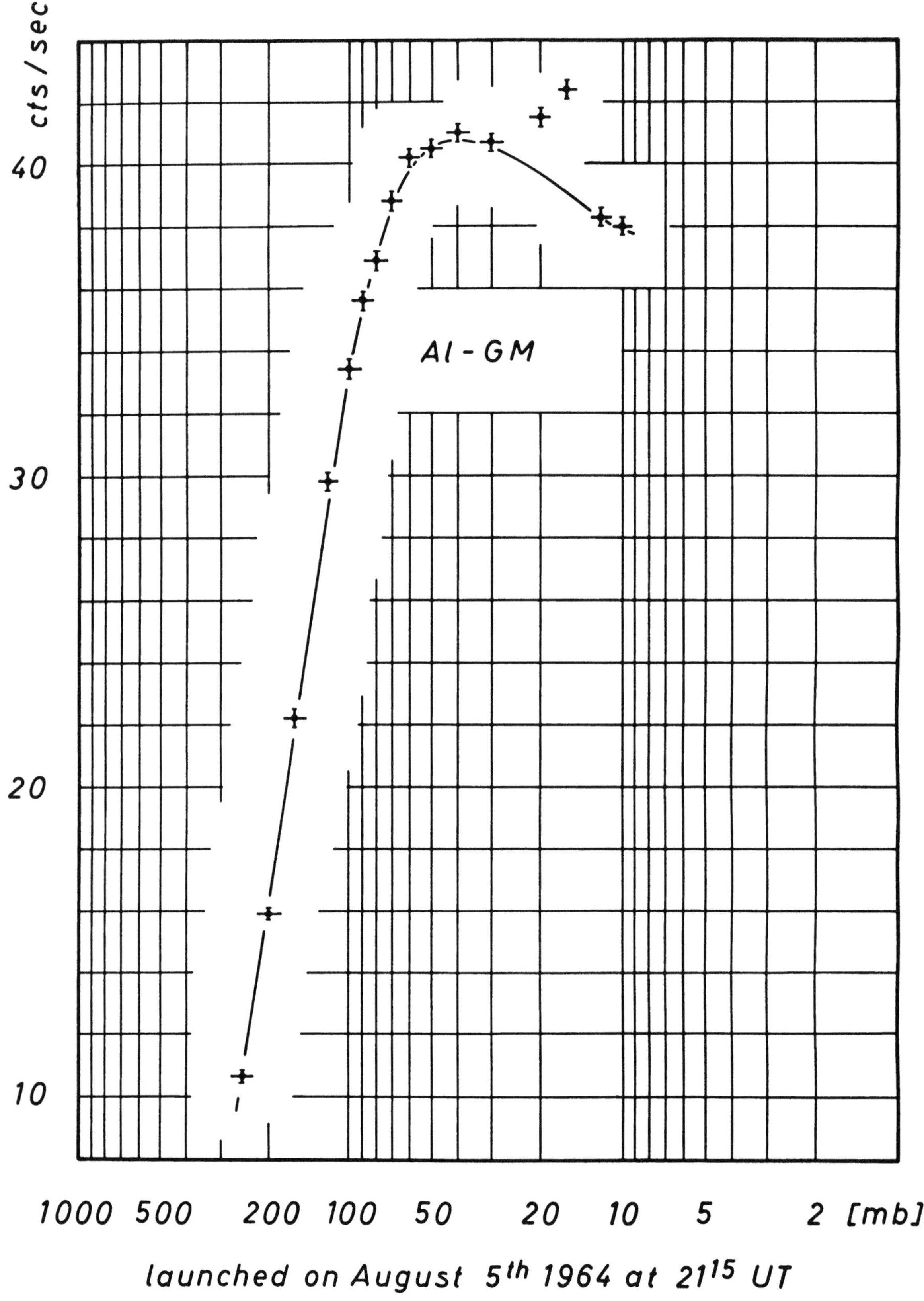

K 12 / 64
cts/sec
40
30
20
10
Al - GM
1000 500 200 100 50 20 10 5 2 [mb]
launched on August 5th 1964 at 21^15 UT

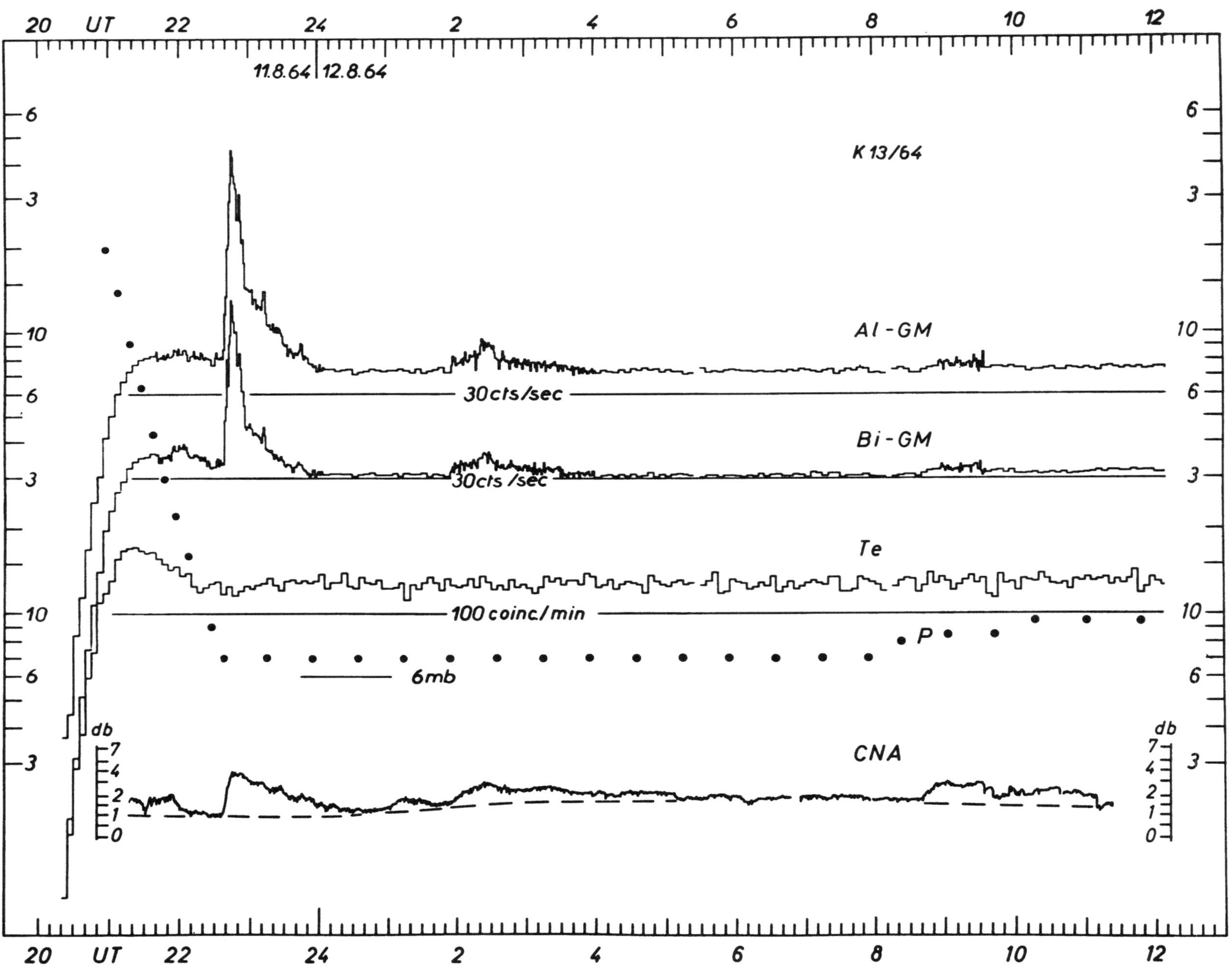

20 UT 22 24 2 4 6 8 10 12
11.8.64 | 12.8.64
K 13/64
Al - GM
30 cts/sec
Bi - GM
30 cts /sec
Te
100 coinc/min
P
6 mb
db
7
4
2
1
0
CNA
20 UT 22 24 2 4 6 8 10 12

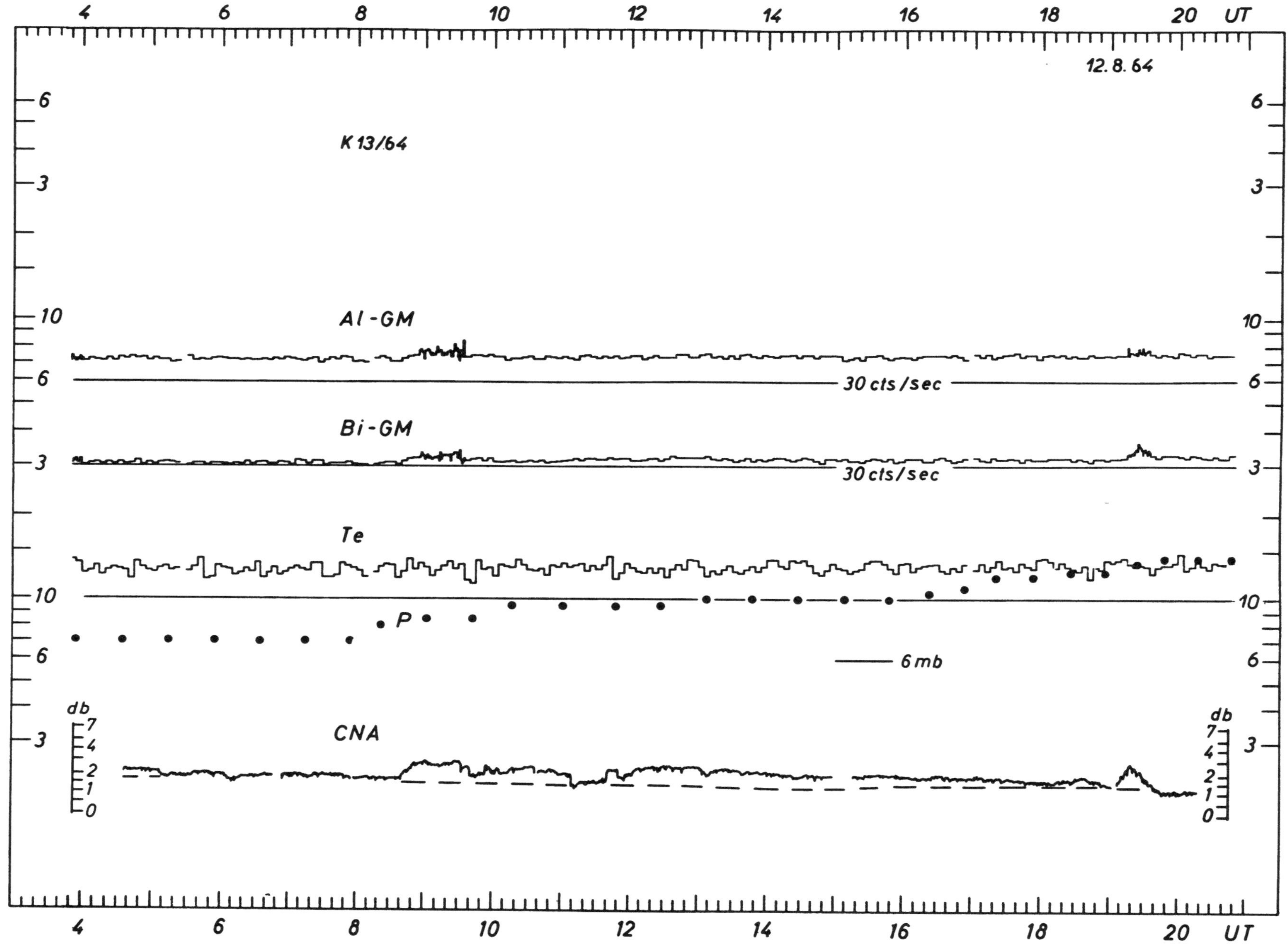
4 6 8 10 12 14 16 18 20 UT
12.8.64
K 13/64
Al-GM
30 cts/sec
Bi-GM
30 cts/sec
Te
P
6 mb
db
7
4
2
1
0
CNA
4 6 8 10 12 14 16 18 20 UT

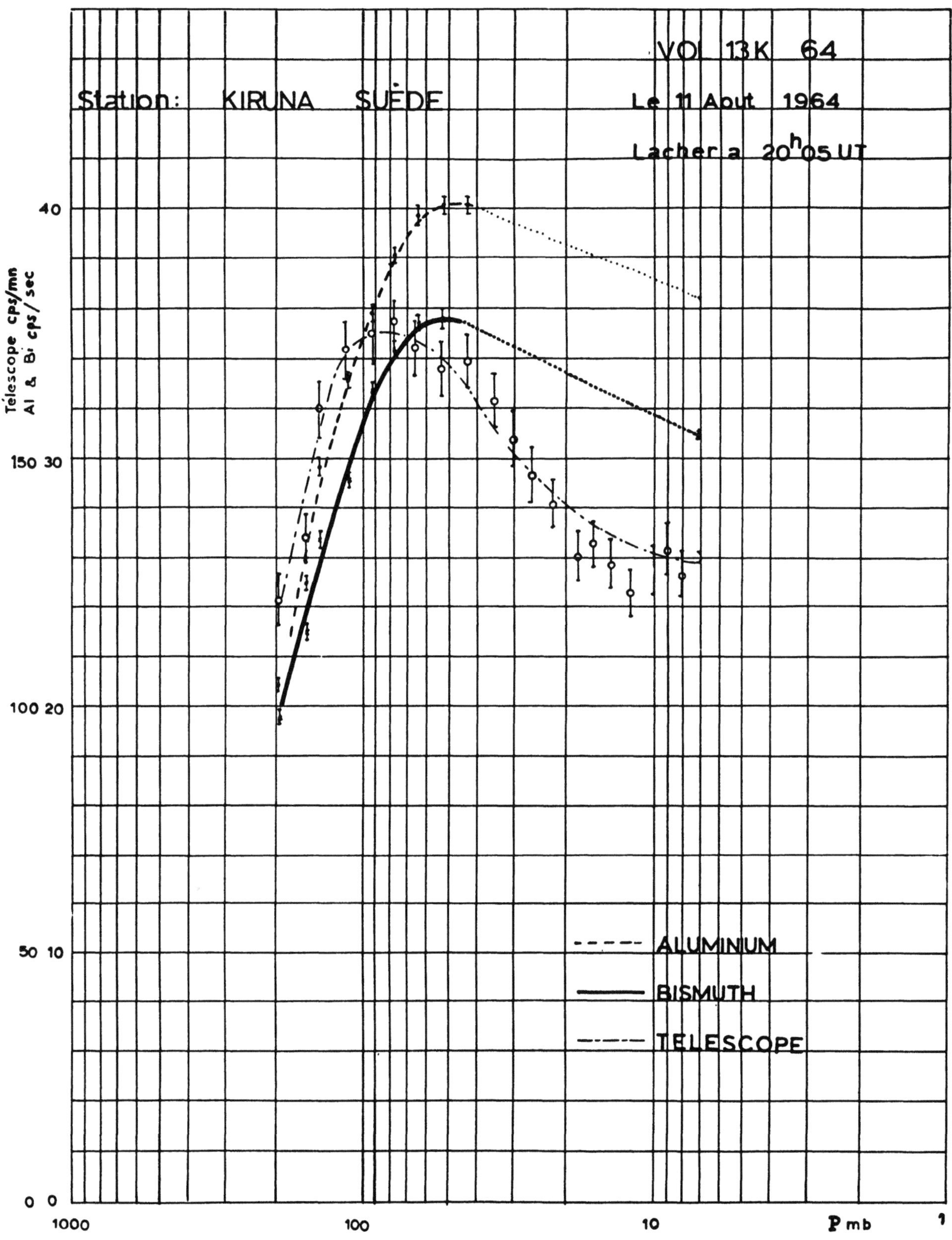

VOL 13K 64
Station: KIRUNA SUÈDE
Le 11 Aout 1964
Lacher a 20h 05 UT
Télescope cps/mn
Al & Bi cps/sec
40
150 30
100 20
50 10
0 0
1000
100
10
P mb
ALUMINIUM
BISMUTH
TELESCOPE

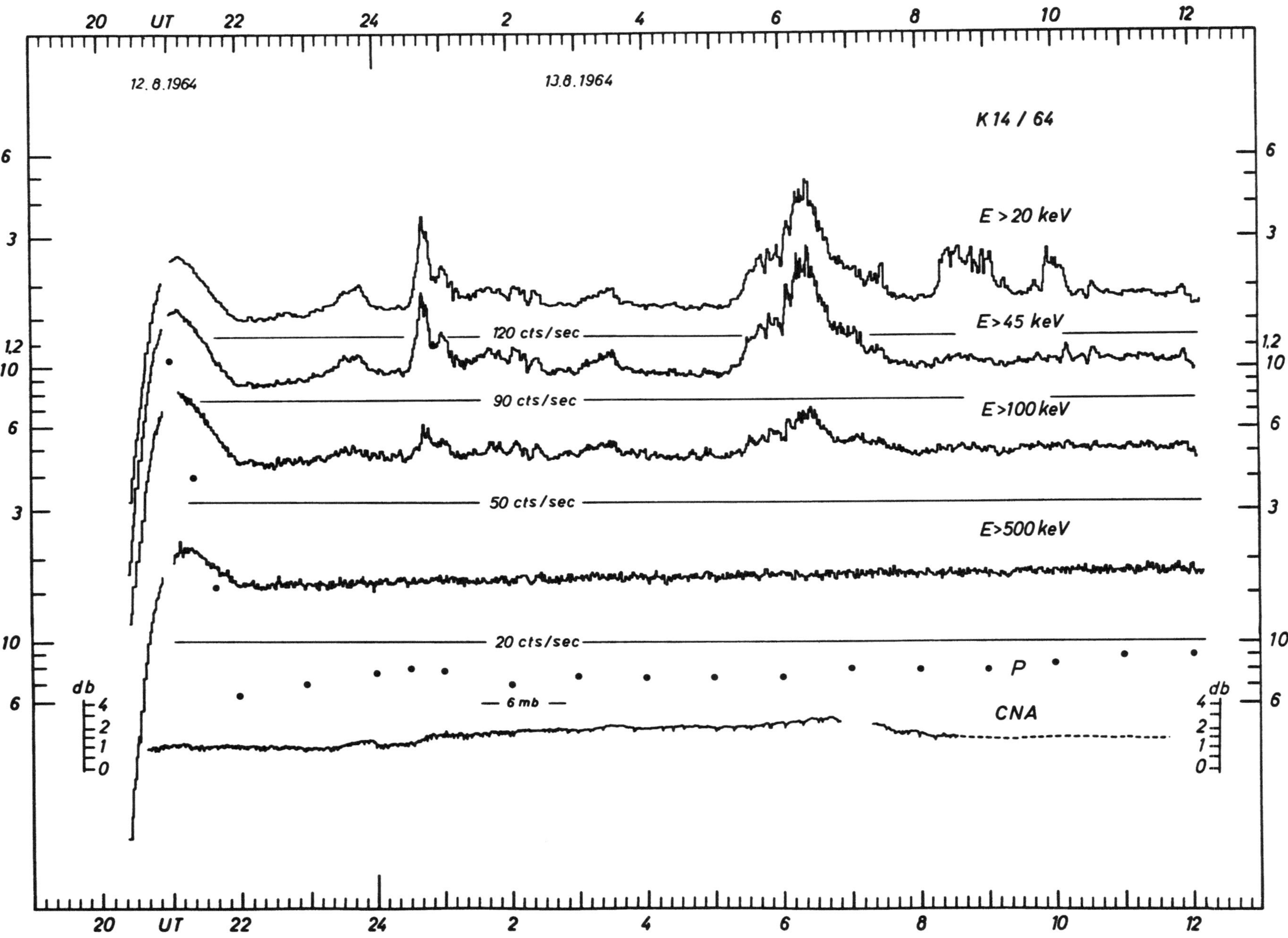

20 UT 22 24 2 4 6 8 10 12
12.8.1964
13.8.1964
K 14 / 64
E > 20 keV
E > 45 keV
E > 100 keV
E > 500 keV
120 cts/sec
90 cts/sec
50 cts/sec
20 cts/sec
6 mb
P
CNA
db
4
2
1
0

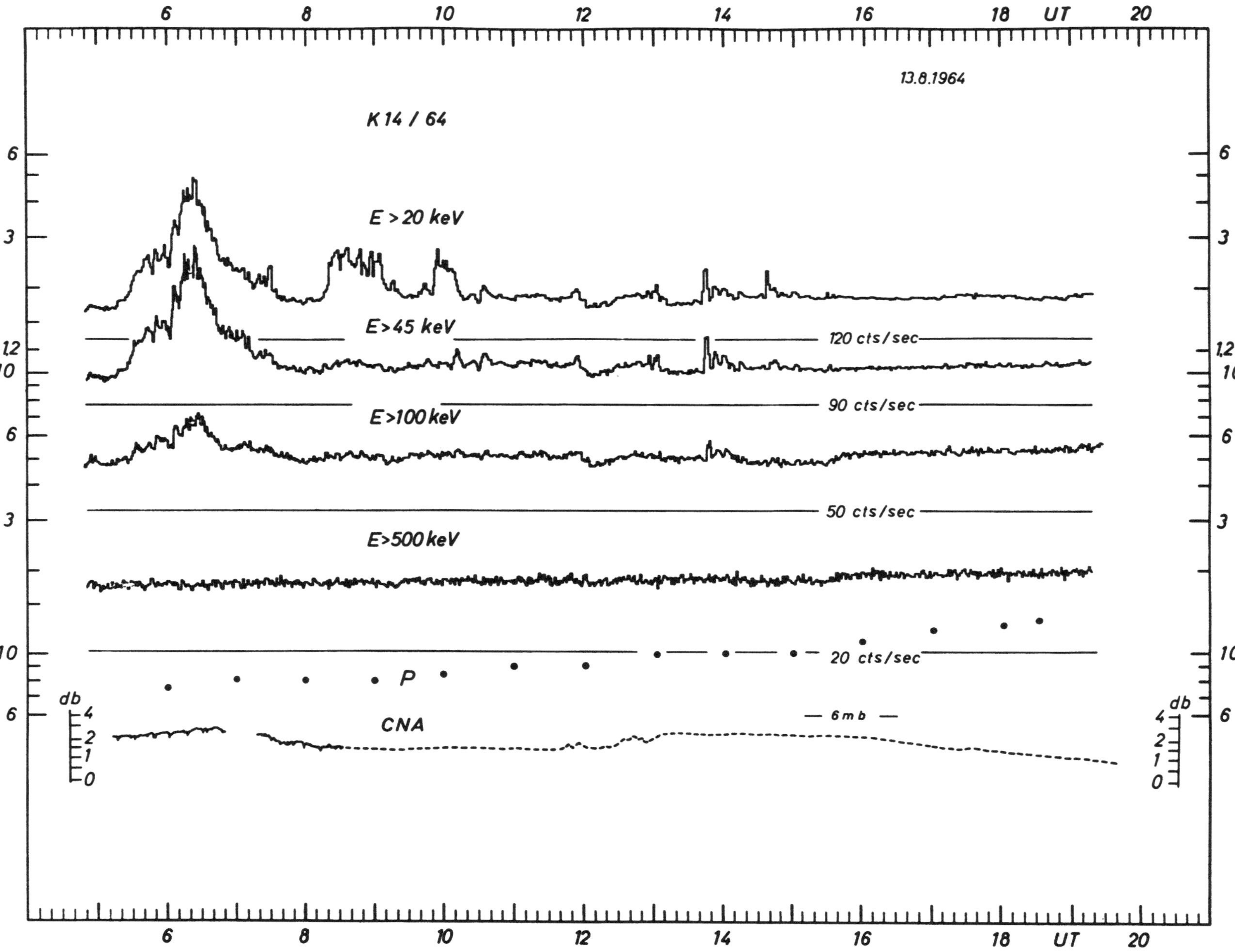

13.8.1964
K 14 / 64
E > 20 keV
E > 45 keV
E > 100 keV
E > 500 keV
120 cts/sec
90 cts/sec
50 cts/sec
20 cts/sec
P
CNA
6 mb
db
UT

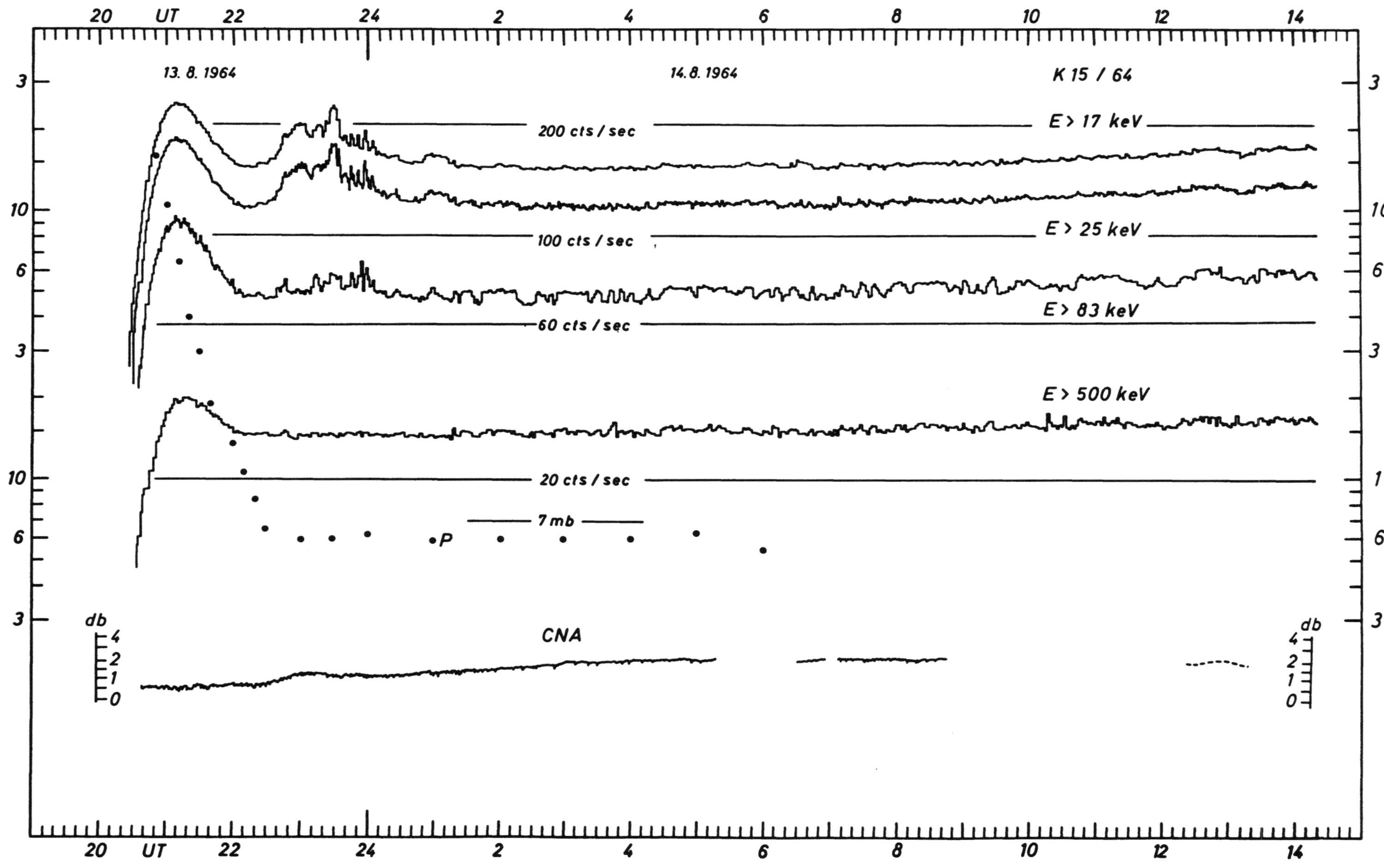

20 UT 22 24 2 4 6 8 10 12 14
13. 8. 1964
14. 8. 1964
K 15 / 64
200 cts / sec
E > 17 keV
100 cts / sec
E > 25 keV
E > 83 keV
60 cts / sec
E > 500 keV
20 cts / sec
7 mb
P
db
4 2 1 0
CNA
db
4 2 1 0
3 10 6 3 10 6 3
20 UT 22 24 2 4 6 8 10 12 14

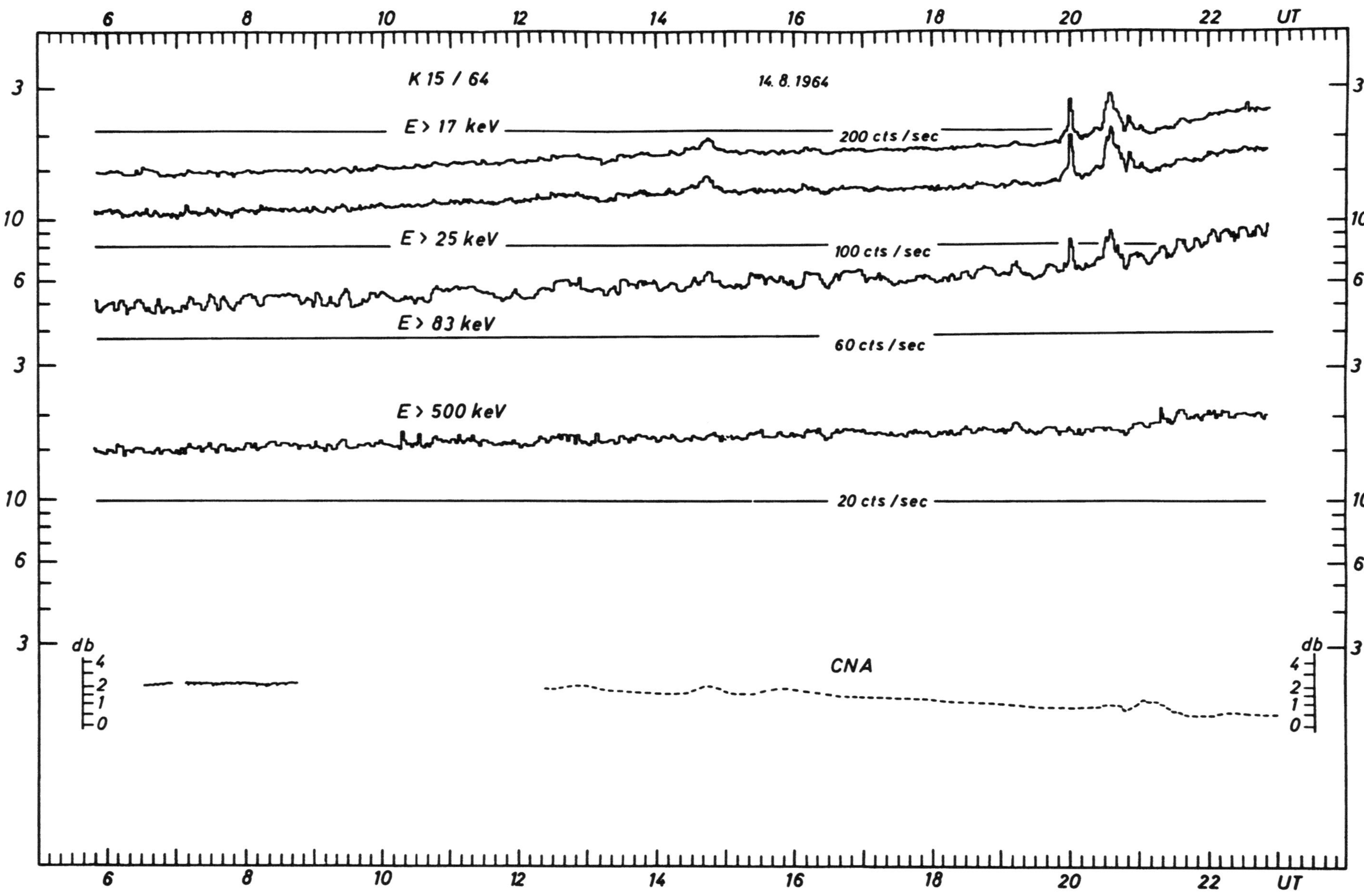

6 8 10 12 14 16 18 20 22 UT
K 15 / 64
14. 8. 1964
3
E > 17 keV
200 cts / sec
3
10
E > 25 keV
100 cts / sec
10
6
E > 83 keV
60 cts / sec
6
3
E > 500 keV
3
10
20 cts / sec
10
6
6
3
db
4
2
1
0
CNA
db
3
4
2
1
0
6 8 10 12 14 16 18 20 22 UT

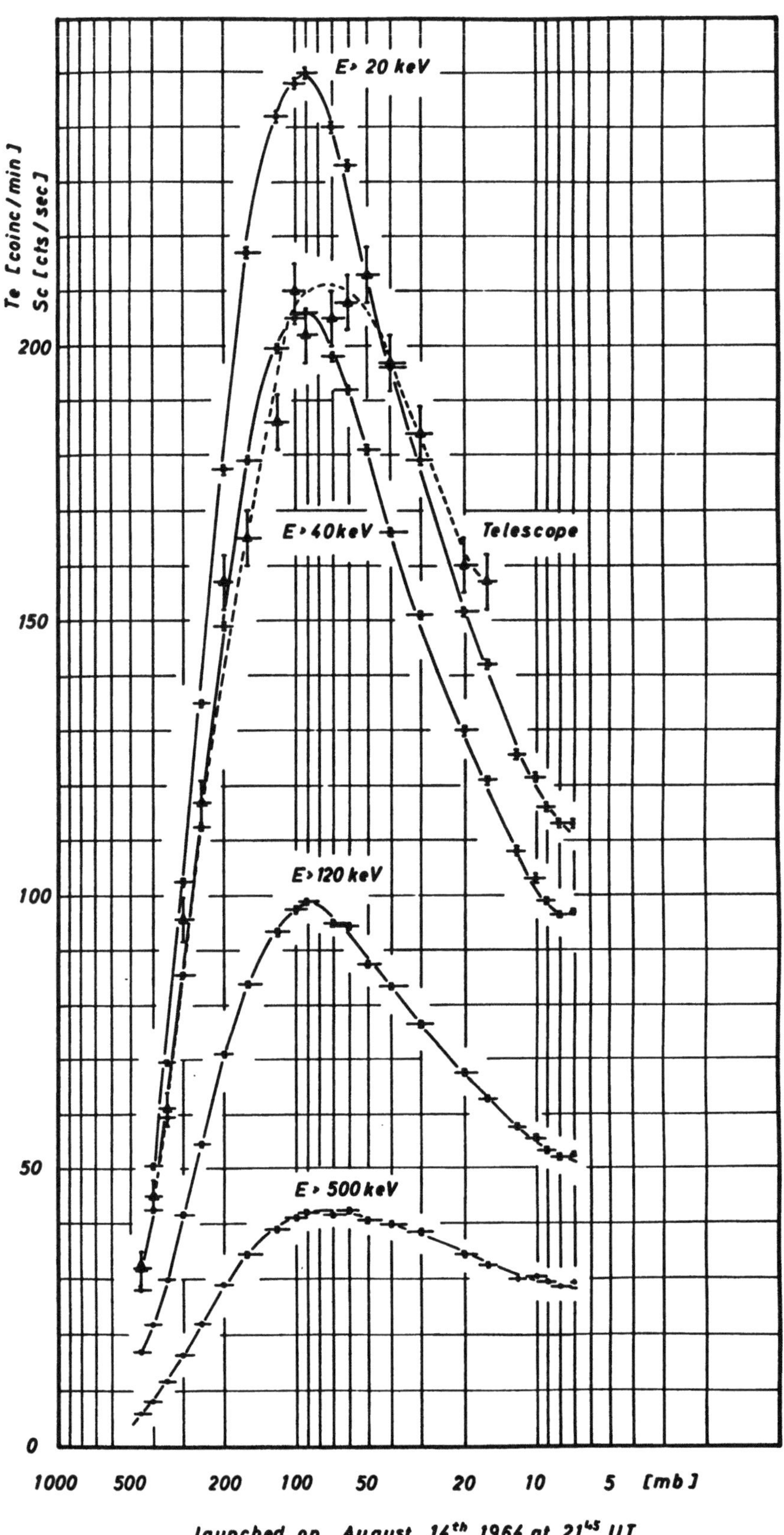

K 16 / 64
Te [coinc/min]
Sc [cts/sec]
E > 20 keV
E > 40 keV
Telescope
E > 120 keV
E > 500 keV
200
150
100
50
0
1000 500 200 100 50 20 10 5 [mb]
launched on August 14th 1964 at 21⁴⁵ UT

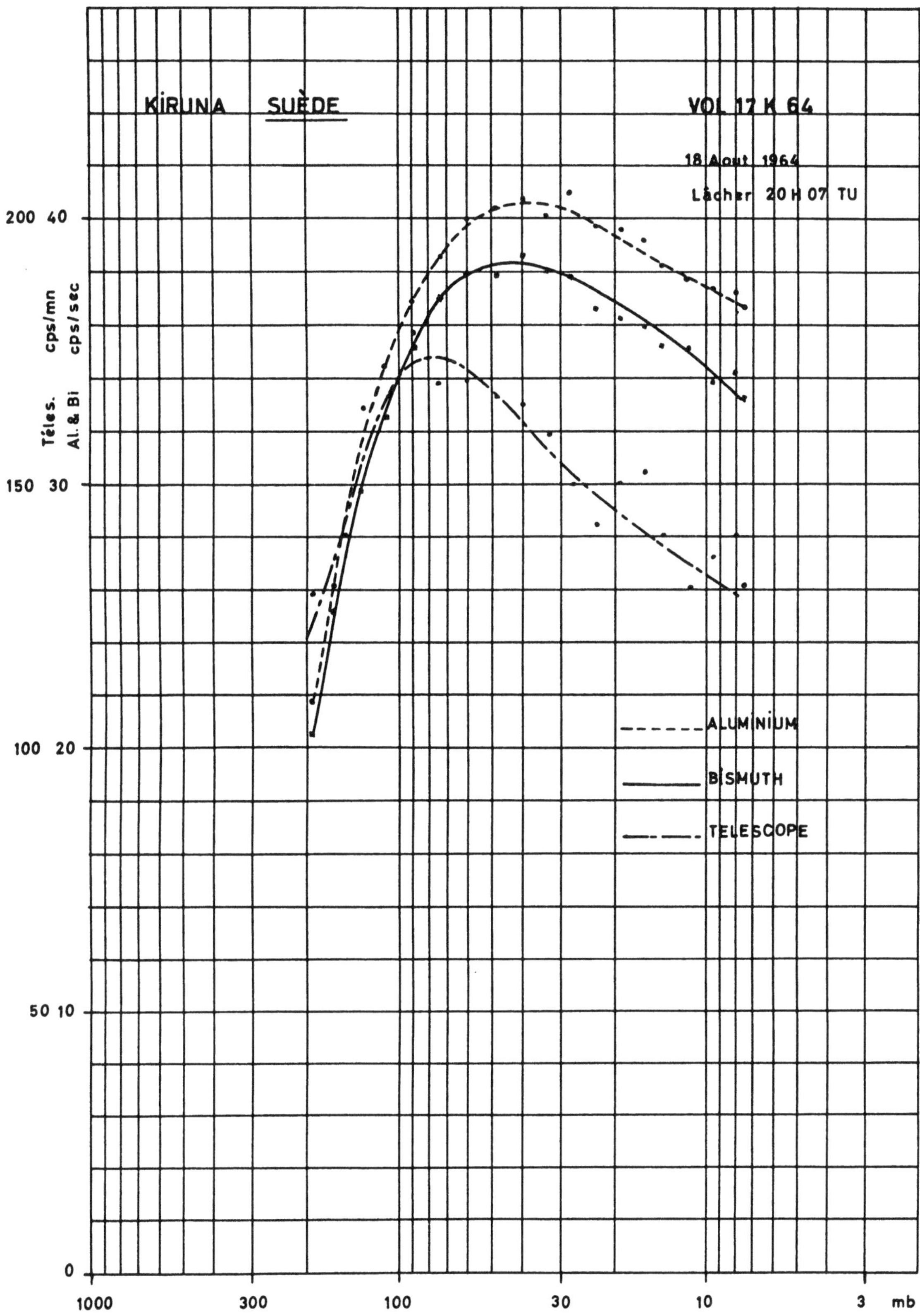

KIRUNA SUÈDE
VOL 17 K 64
18 Aout 1964
Lâcher 20 H 07 TU
cps/mn
cps/sec
Téles.
Al. & Bi
200 40
150 30
100 20
50 10
0
1000
300
100
30
10
3   mb
ALUMINIUM
BISMUTH
TELESCOPE

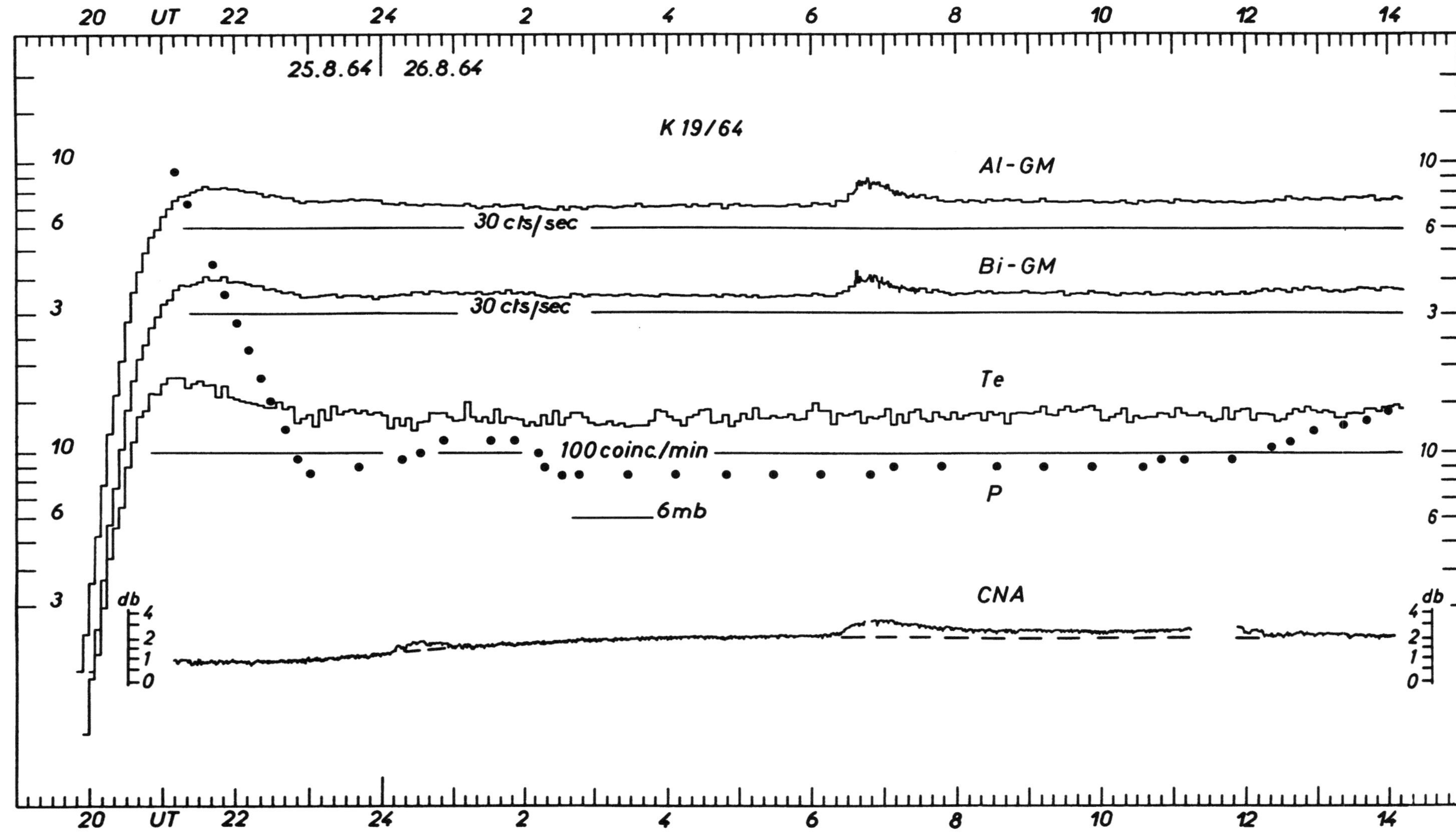

25.8.64 | 26.8.64
K 19/64
Al-GM
30 cts/sec
Bi-GM
30 cts/sec
Te
100 coinc./min
6mb
P
db
CNA
UT

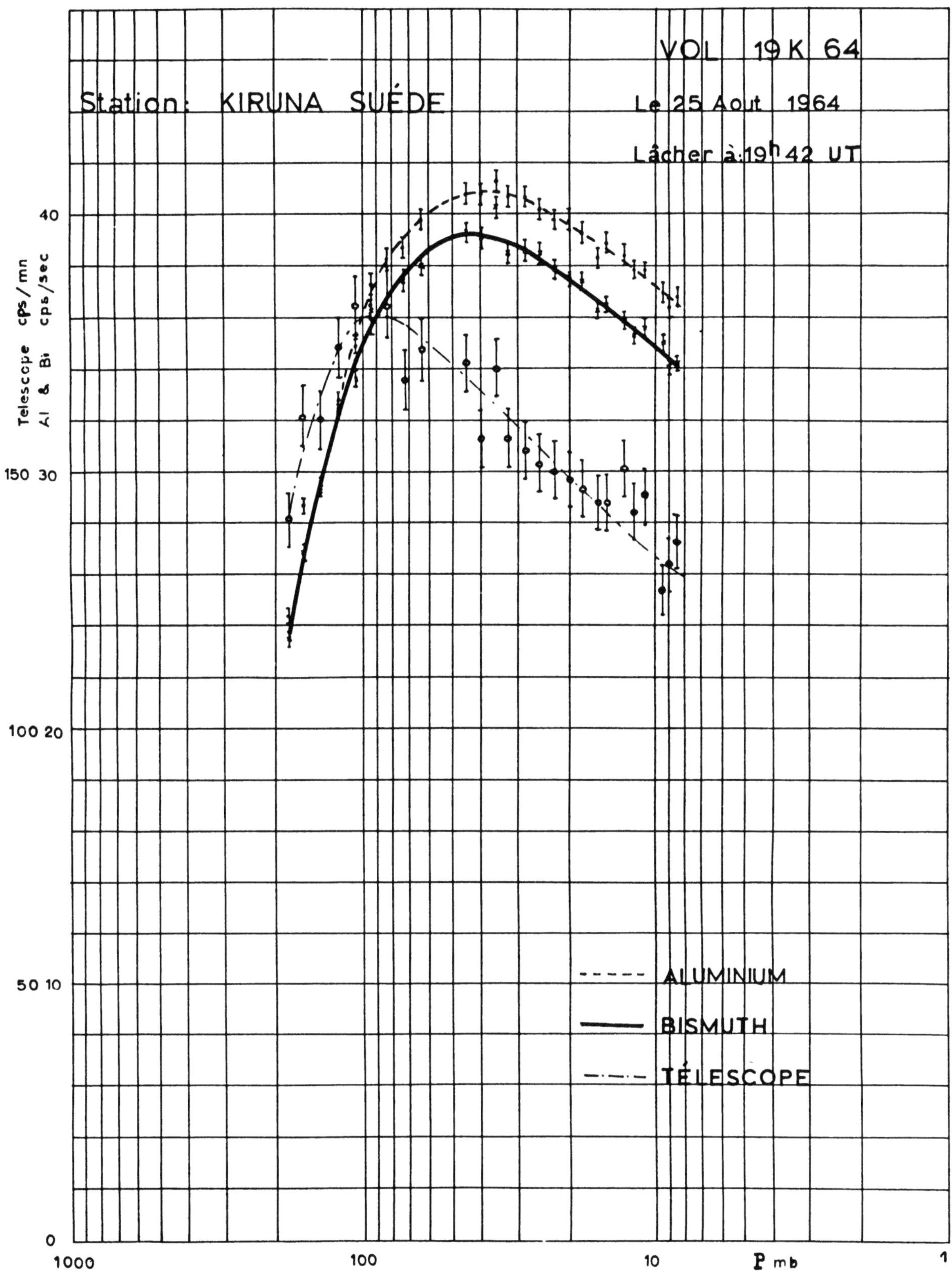

VOL 19 K 64
Station: KIRUNA SUÉDE
Le 25 Aout 1964
Lâcher à 19h 42 UT
Telescope cps/mn
Al & Bi cps/sec
40
30
20
10
150
100
50
0
1000
100
10
1
P mb
----- ALUMINIUM
BISMUTH
-.-.- TÉLESCOPE

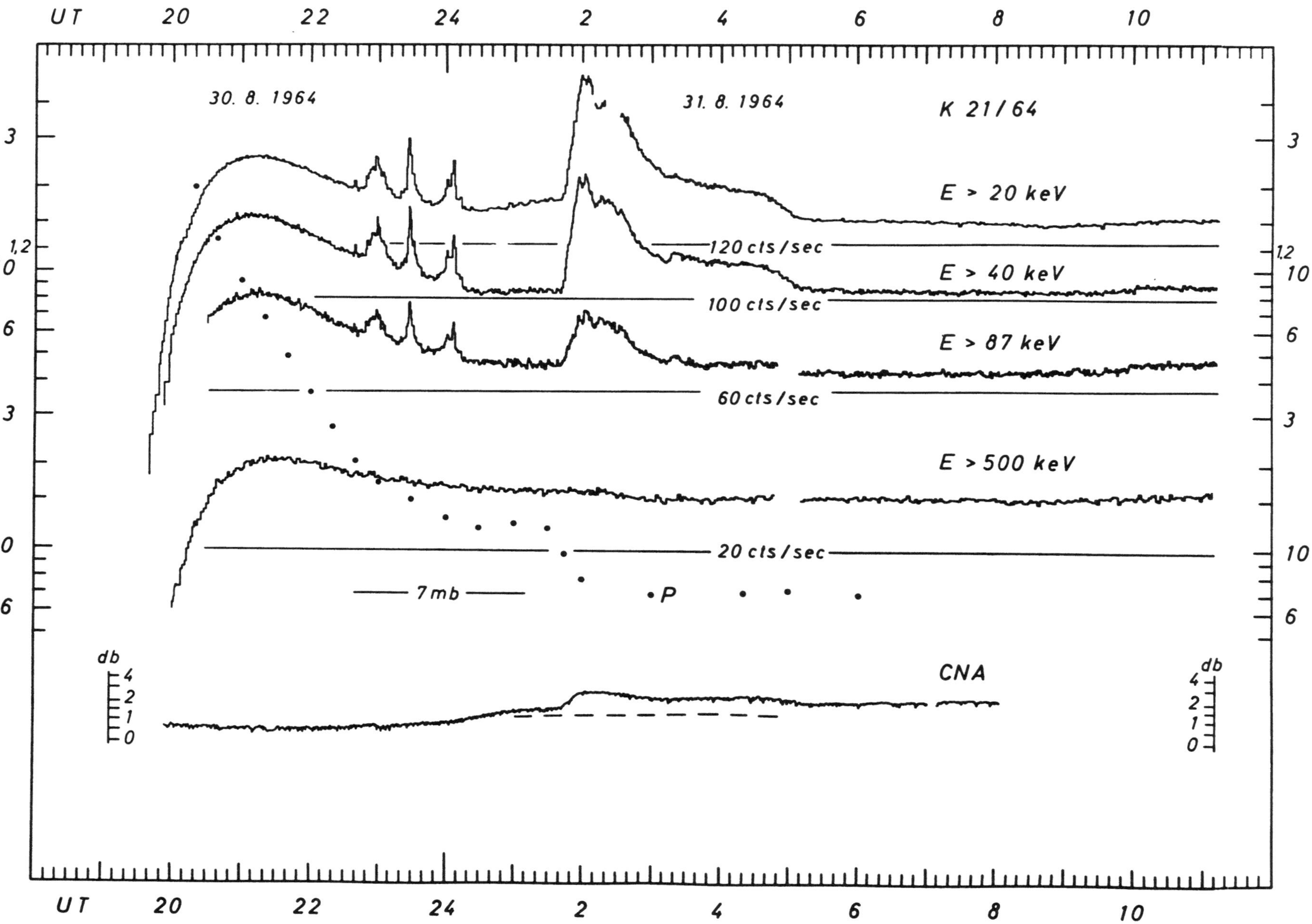

UT
30. 8. 1964
31. 8. 1964
K 21/64
E > 20 keV
120 cts/sec
E > 40 keV
100 cts/sec
E > 87 keV
60 cts/sec
E > 500 keV
20 cts/sec
7 mb
P
CNA
db

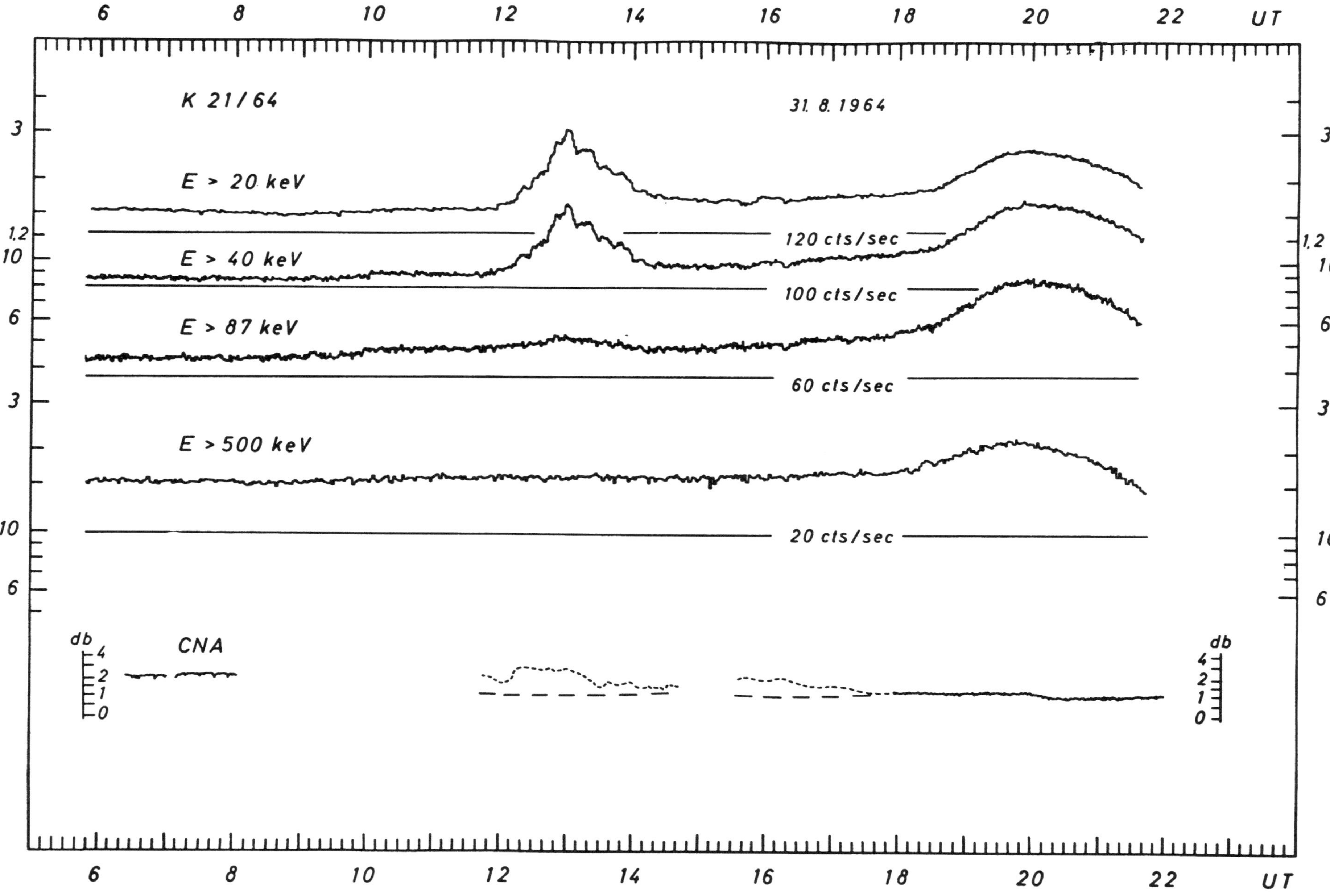
6 8 10 12 14 16 18 20 22 UT
K 21/64
31. 8. 1964
E > 20 keV
120 cts/sec
E > 40 keV
100 cts/sec
E > 87 keV
60 cts/sec
E > 500 keV
20 cts/sec
db 4 2 1 0
CNA
db 4 2 1 0
6 8 10 12 14 16 18 20 22 UT

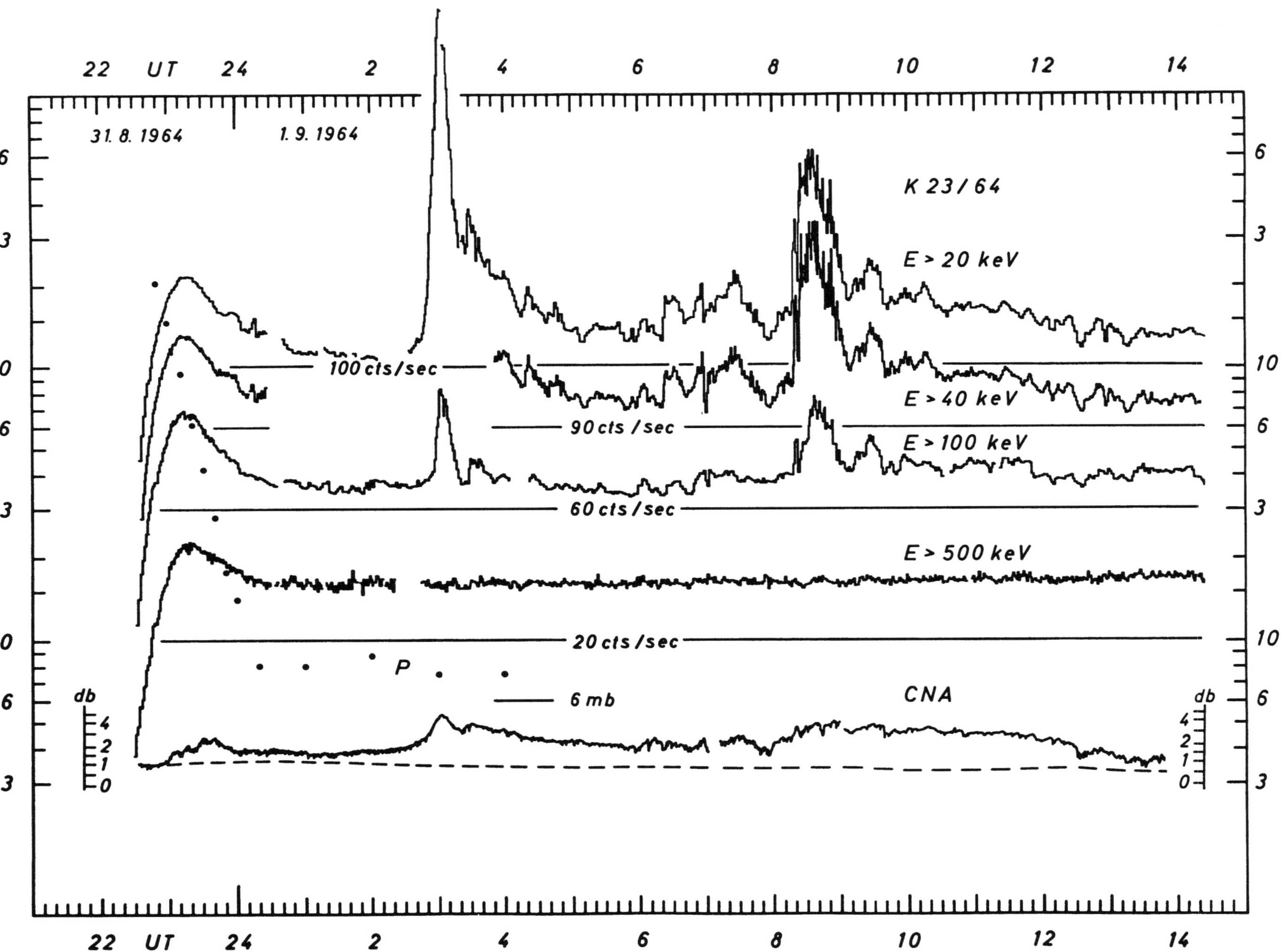

22 UT 24 2 4 6 8 10 12 14
31. 8. 1964
1. 9. 1964
K 23 / 64
E > 20 keV
100 cts/sec
E > 40 keV
90 cts / sec
E > 100 keV
60 cts/sec
E > 500 keV
20 cts/sec
P
6 mb
CNA
db
4
2
1
0

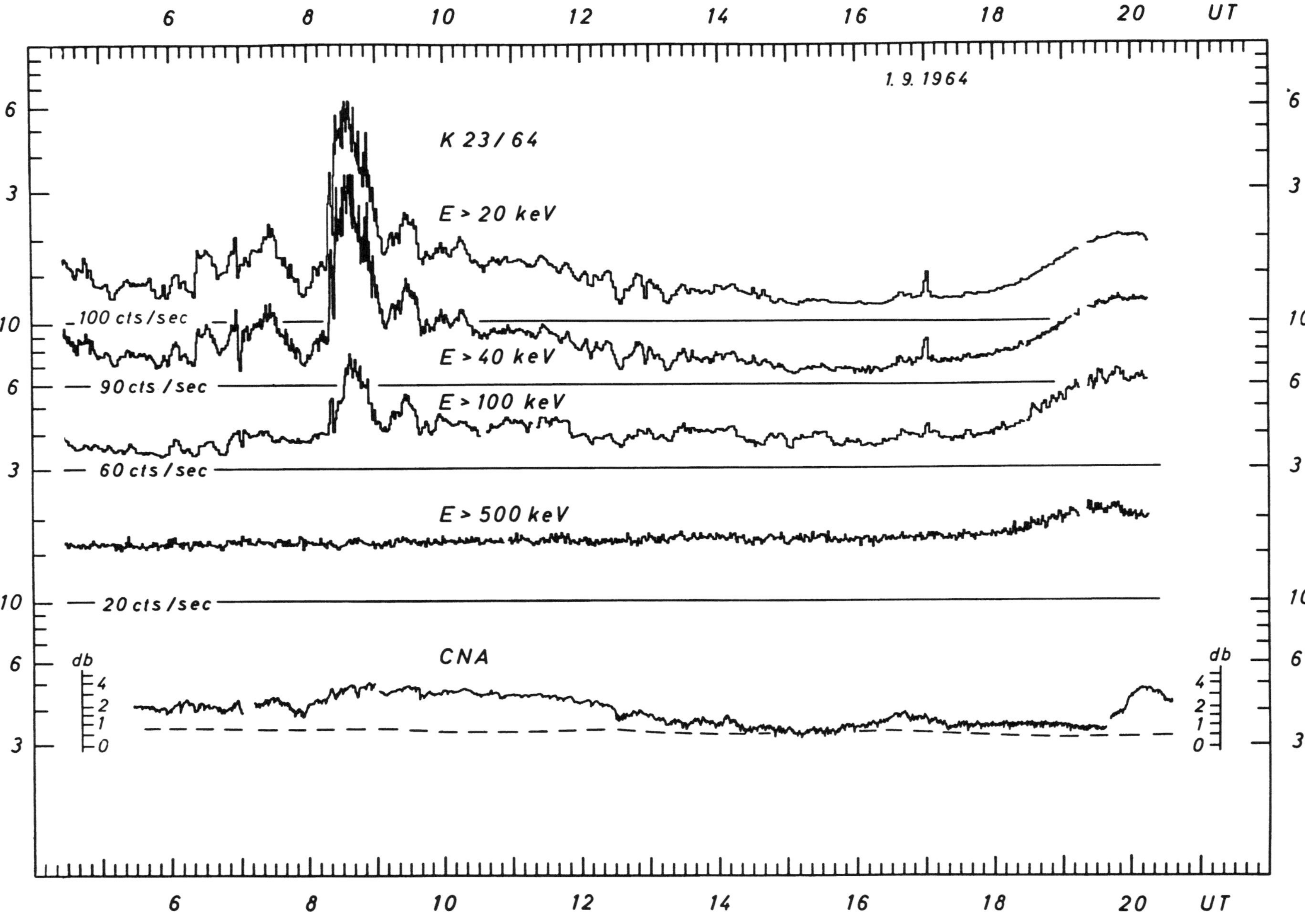

1.9.1964
K 23/64
E > 20 keV
E > 40 keV
E > 100 keV
E > 500 keV
CNA
100 cts/sec
90 cts/sec
60 cts/sec
20 cts/sec
db
4
2
1
0
UT

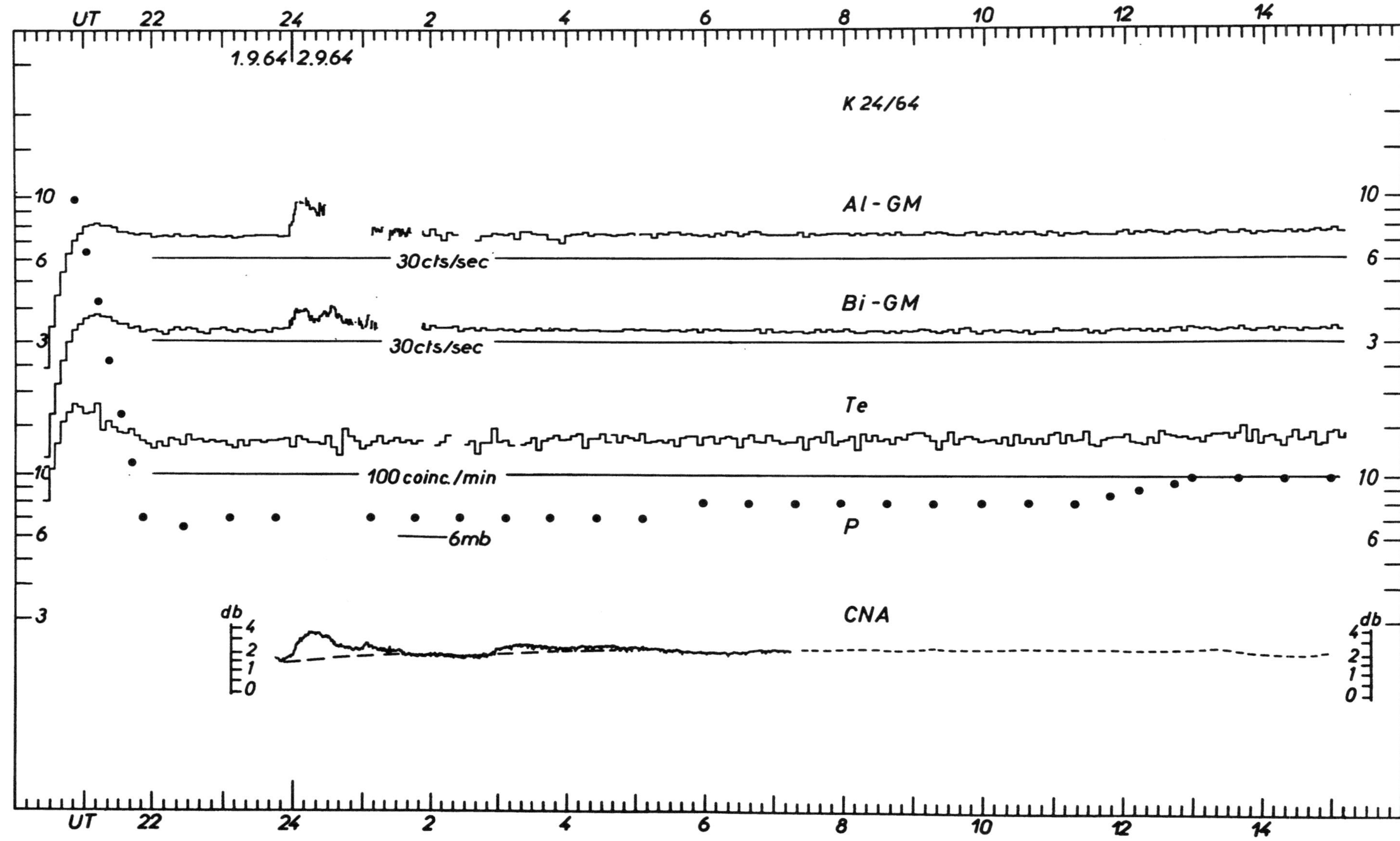

UT 22 24 2 4 6 8 10 12 14
1.9.64 | 2.9.64
K 24/64
Al - GM
30 cts/sec
Bi - GM
30 cts/sec
Te
100 coinc./min
6mb
P
CNA
db 4 2 1 0
UT 22 24 2 4 6 8 10 12 14

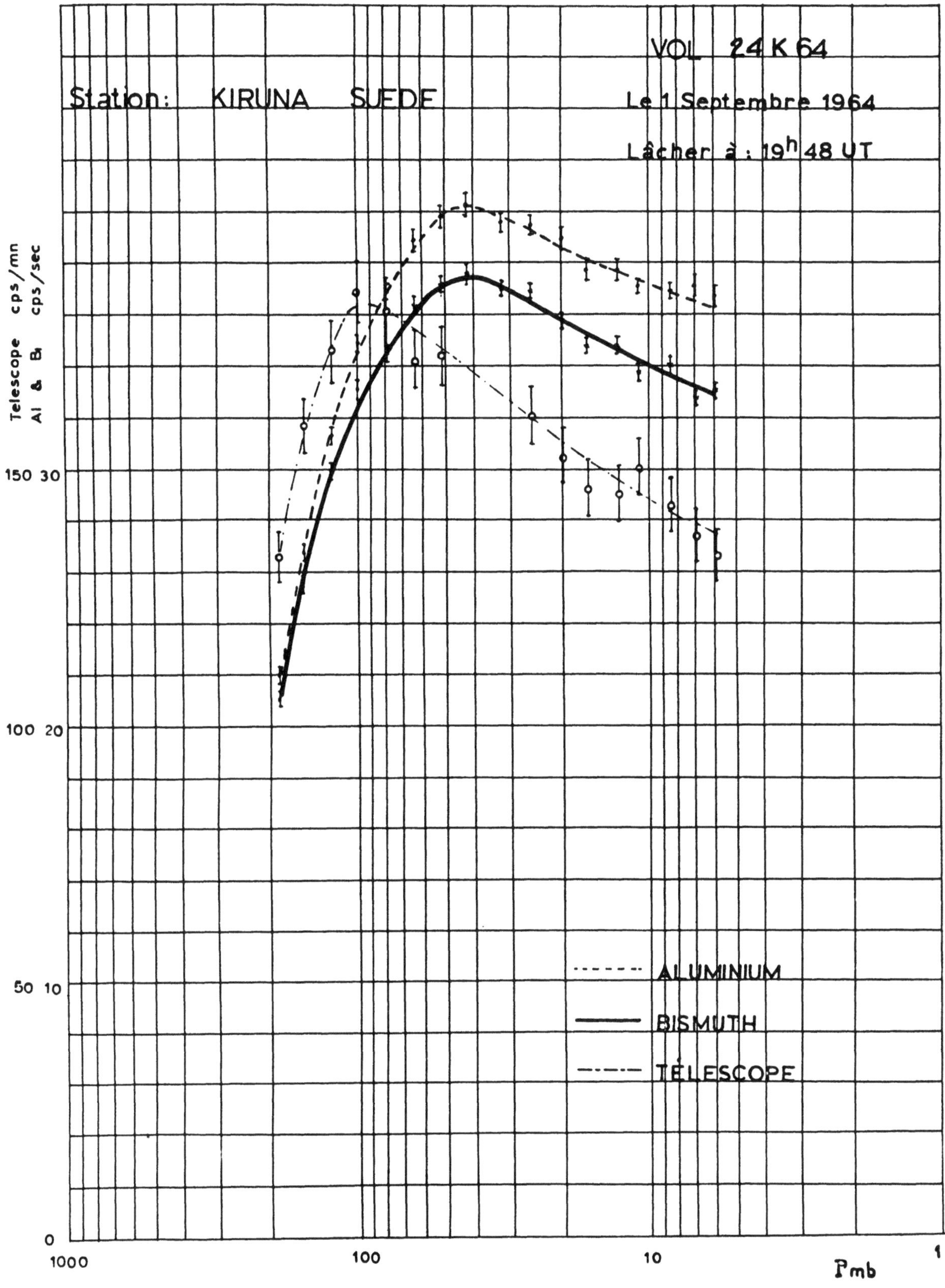

VOL 24 K 64
Station: KIRUNA SUEDE
Le 1 Septembre 1964
Lâcher à : 19h 48 UT
Telescope cps/mn
Al & Bi cps/sec
150 30
100 20
50 10
0
1000
100
10
1
Pmb
ALUMINIUM
BISMUTH
TELESCOPE

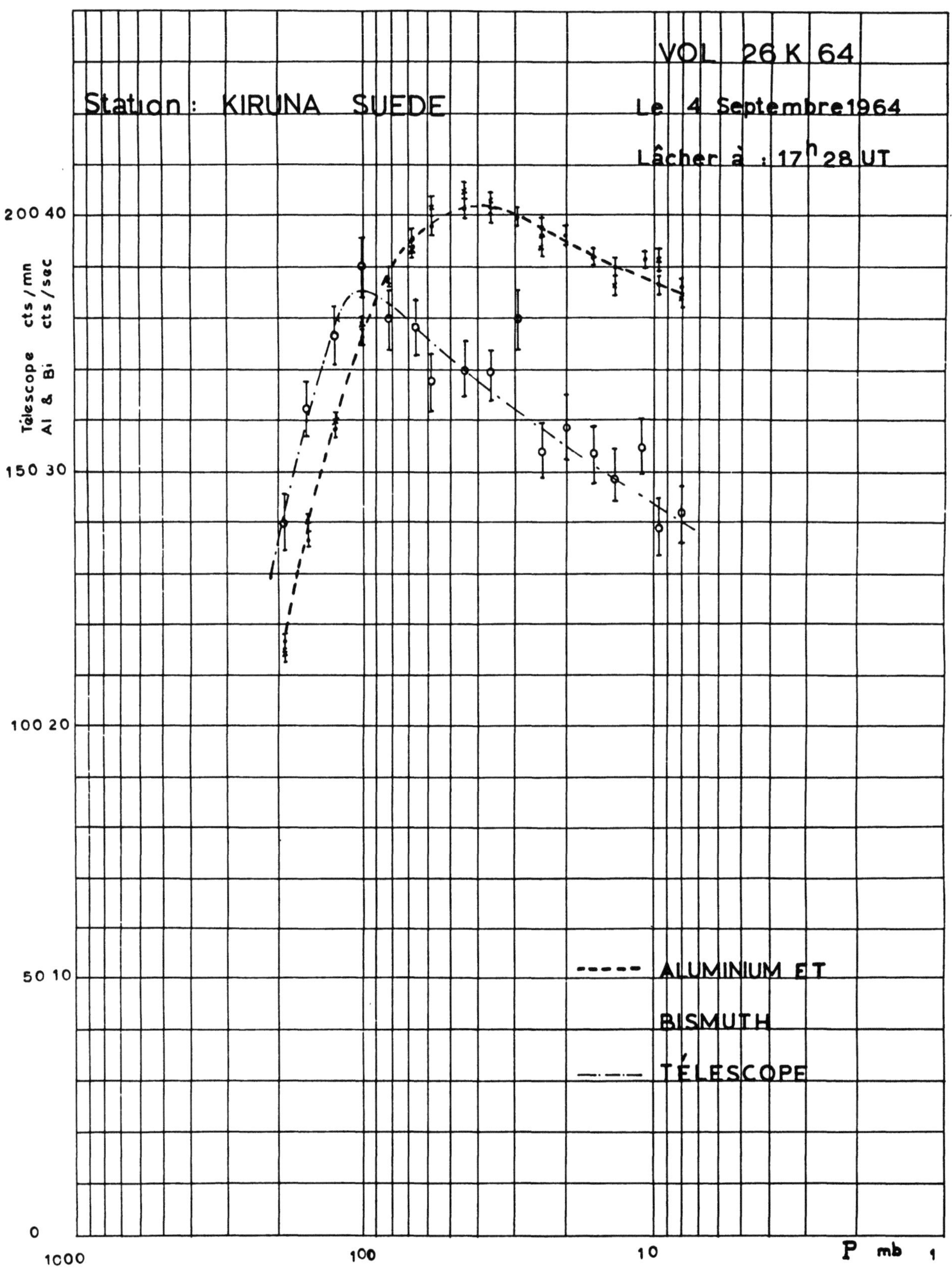

VOL 26 K 64
Station : KIRUNA SUEDE
Le 4 Septembre 1964
Lâcher à : 17h 28 UT
Télescope cts/mn
Al & Bi cts/sec
200 40
150 30
100 20
50 10
0
1000
100
10
P mb
----- ALUMINIUM ET BISMUTH
-·-·- TÉLESCOPE

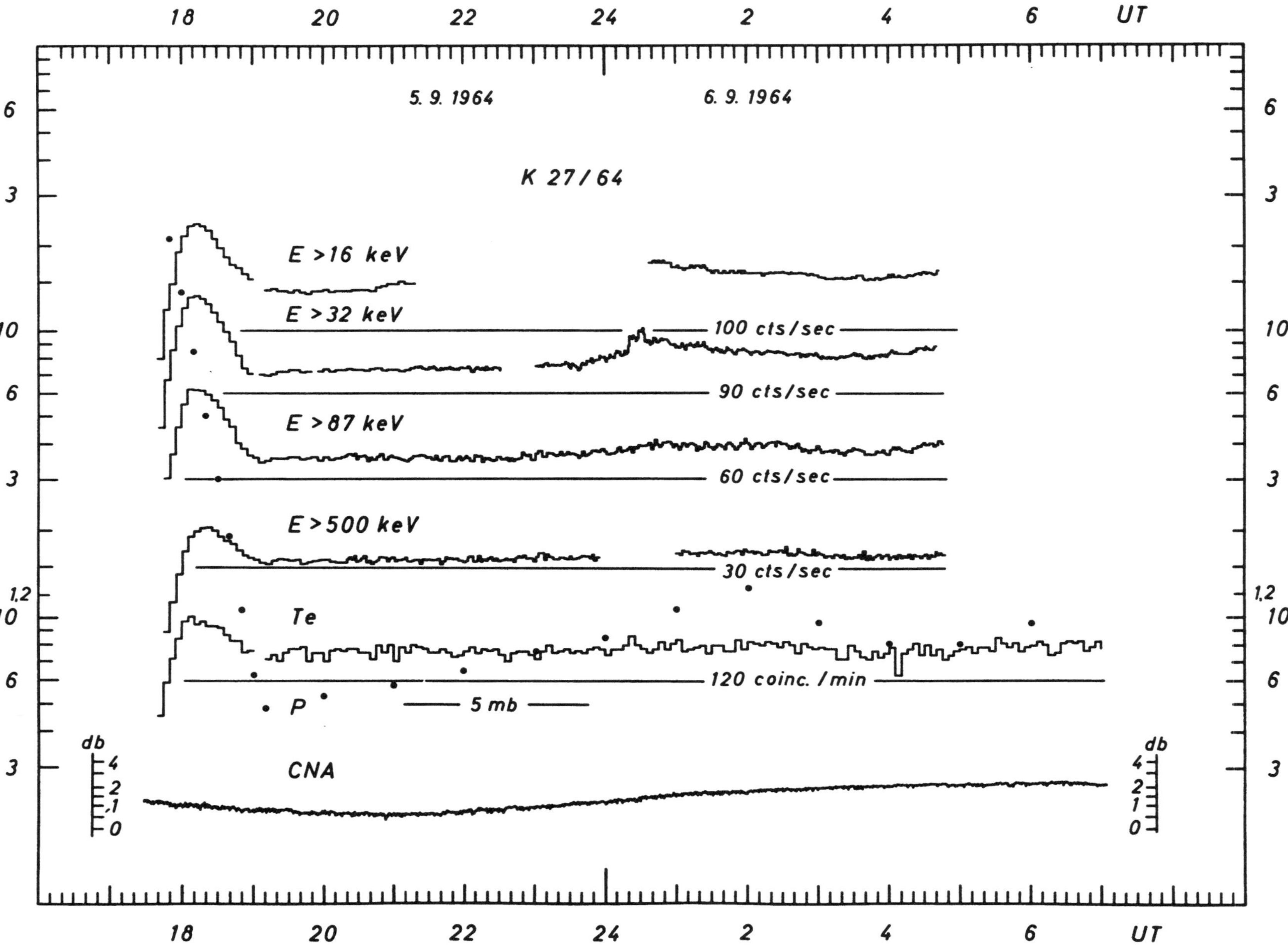

UT
5. 9. 1964
6. 9. 1964
K 27/64
E > 16 keV
E > 32 keV
100 cts/sec
90 cts/sec
E > 87 keV
60 cts/sec
E > 500 keV
30 cts/sec
Te
120 coinc./min
P
5 mb
db
4
2
1
0
CNA
UT

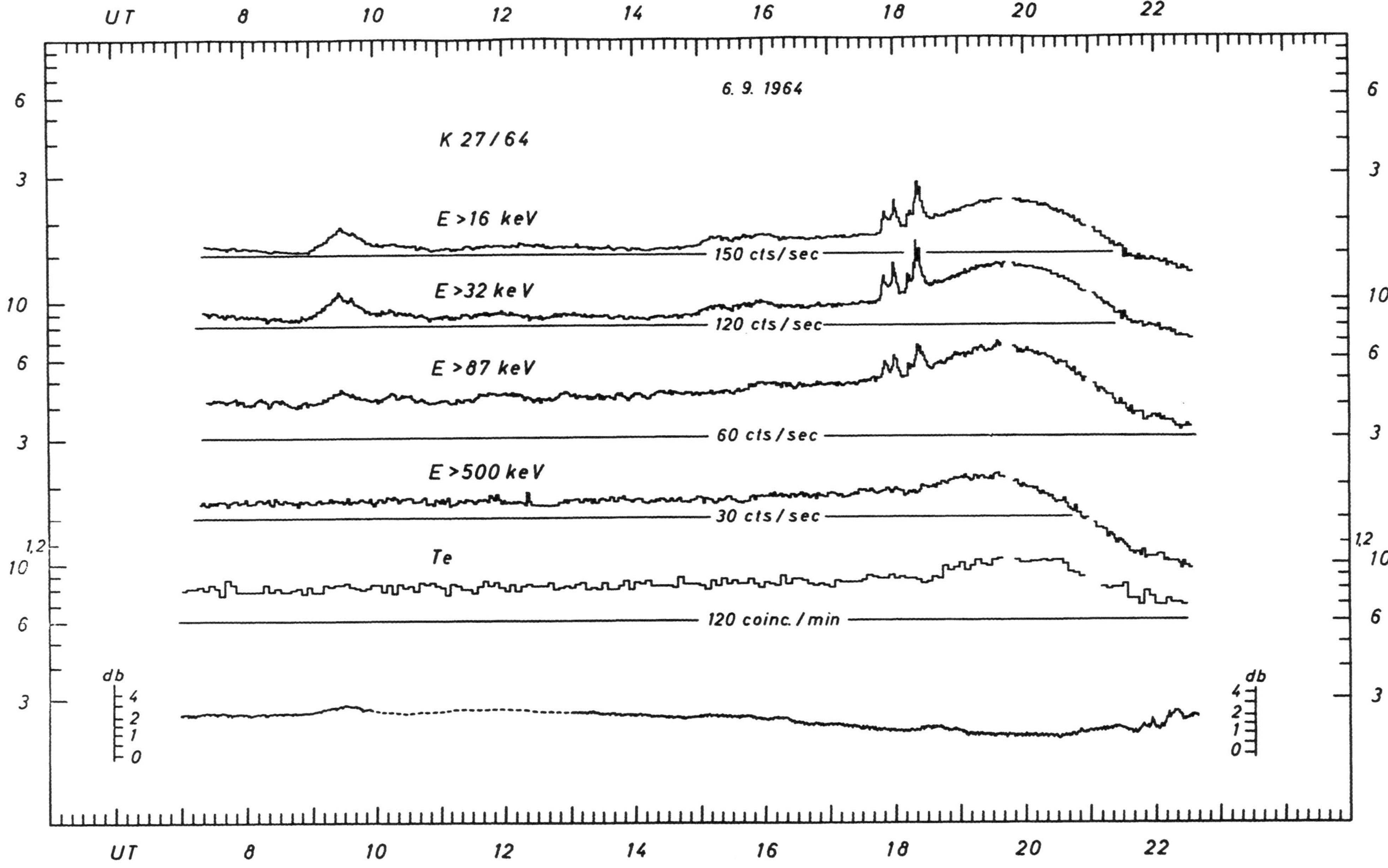
UT
6. 9. 1964
K 27/64
E > 16 keV
150 cts/sec
E > 32 keV
120 cts/sec
E > 87 keV
60 cts/sec
E > 500 keV
30 cts/sec
Te
120 coinc. / min
db
4
2
1
0
UT

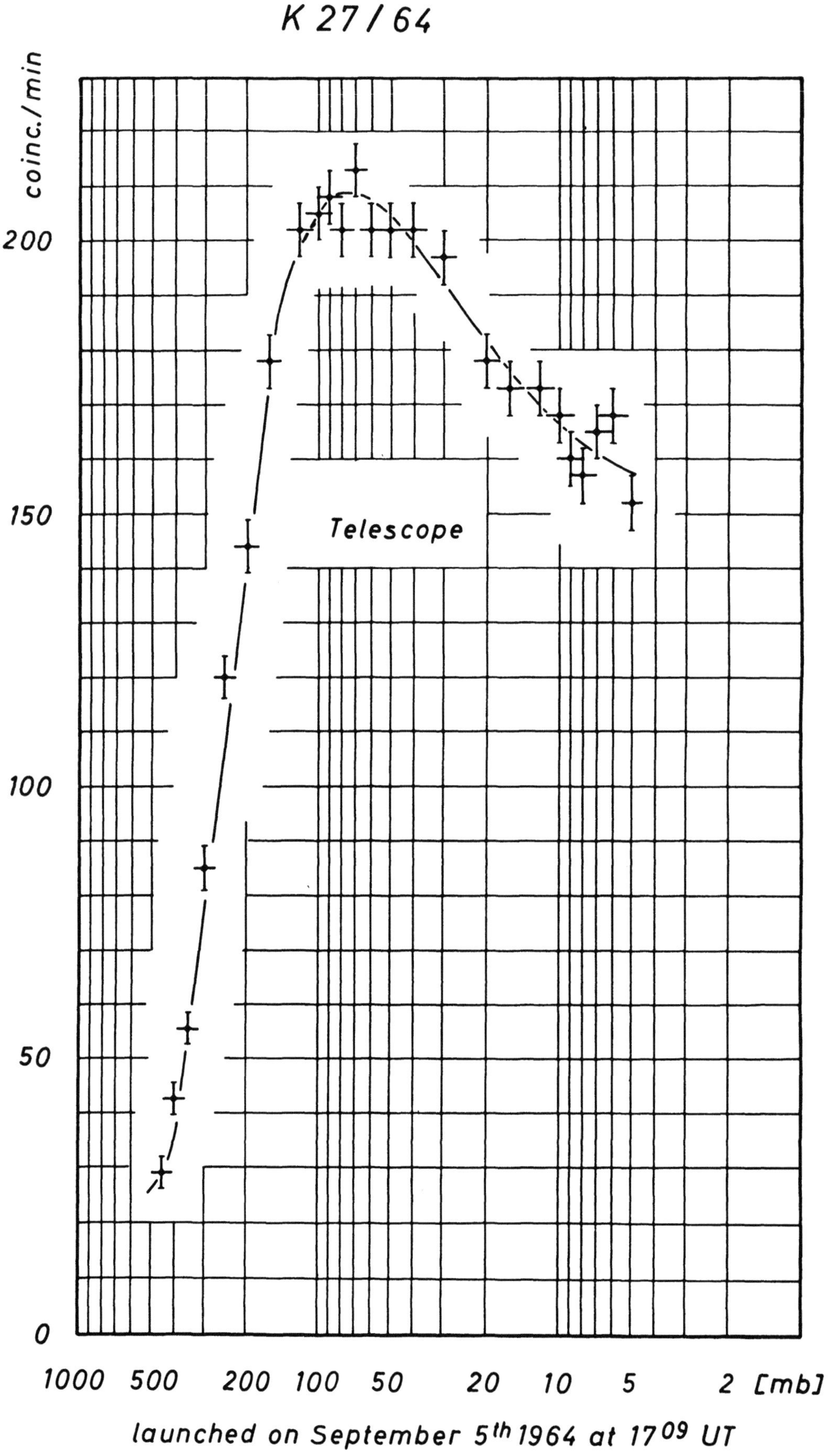

K 27 / 64
coinc./min
200
150
100
50
0
Telescope
1000 500 200 100 50 20 10 5 2 [mb]
launched on September 5th 1964 at 17 09 UT

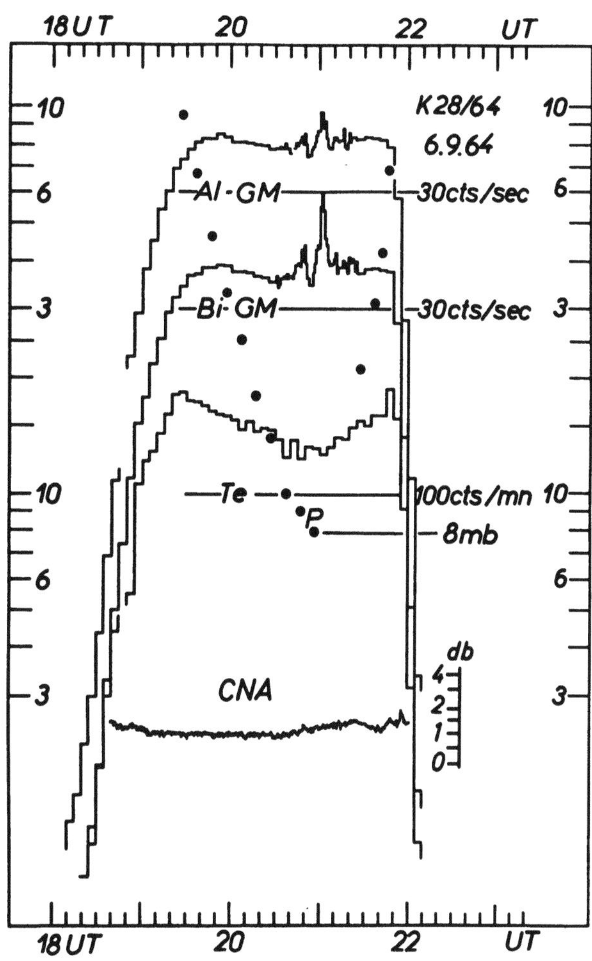

18 U T
20
22
UT
10
K28/64
6.9.64
6
Al · GM
30cts/sec
6
3
Bi · GM
30cts/sec
3
10
Te
100cts/mn
10
P
8mb
6
6
db
4
2
CNA
3
1
3
0
18 UT
20
22
UT

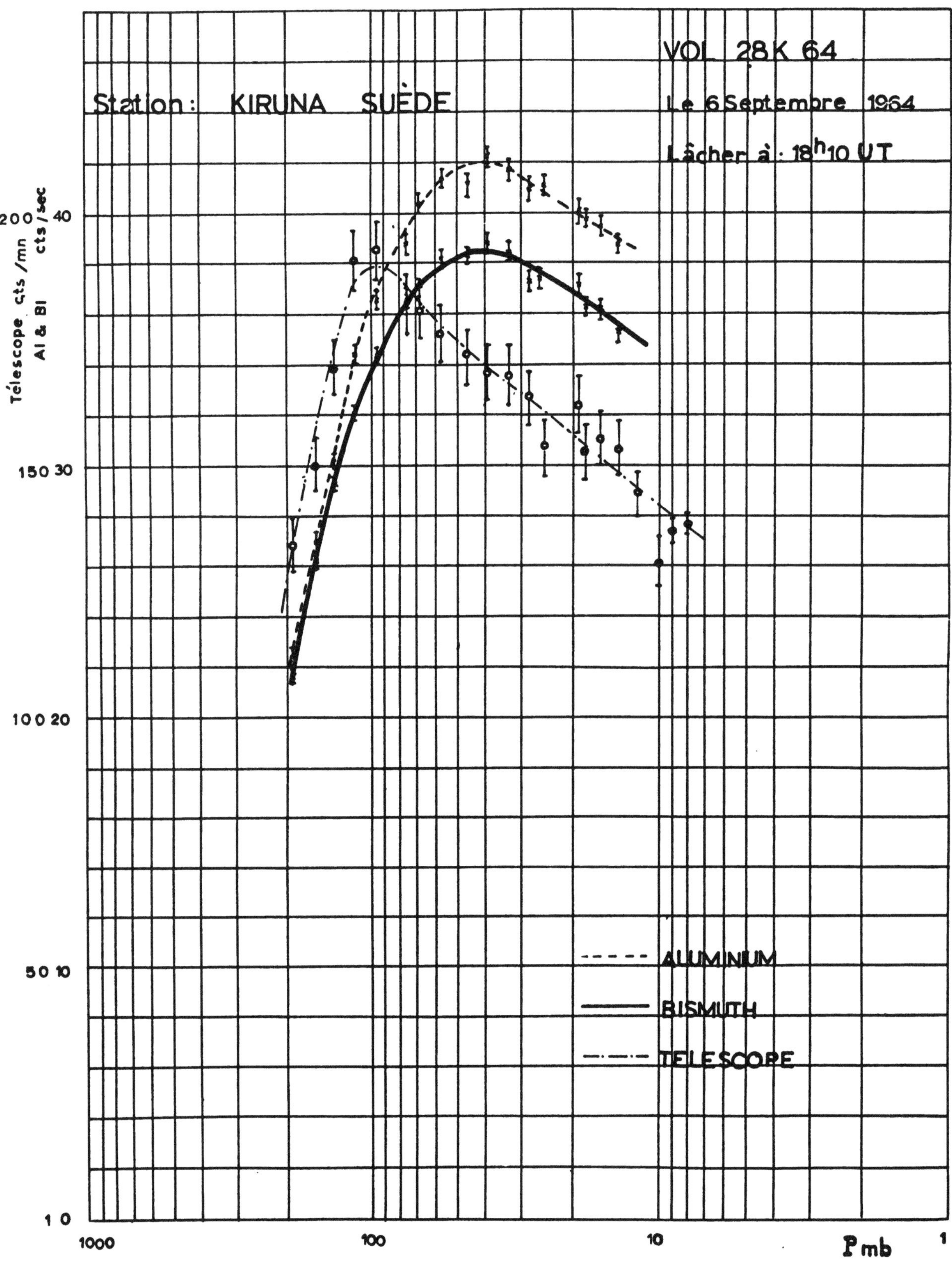

VOL 28K 64
Station : KIRUNA SUÈDE
Le 6 Septembre 1964
Lâcher à : 18h10 UT
Télescope cts /mn
Al & Bi
cts /sec
200
40
150 30
100 20
50 10
1 0
1000
100
10
1
P mb
ALUMINIUM
BISMUTH
TELESCOPE

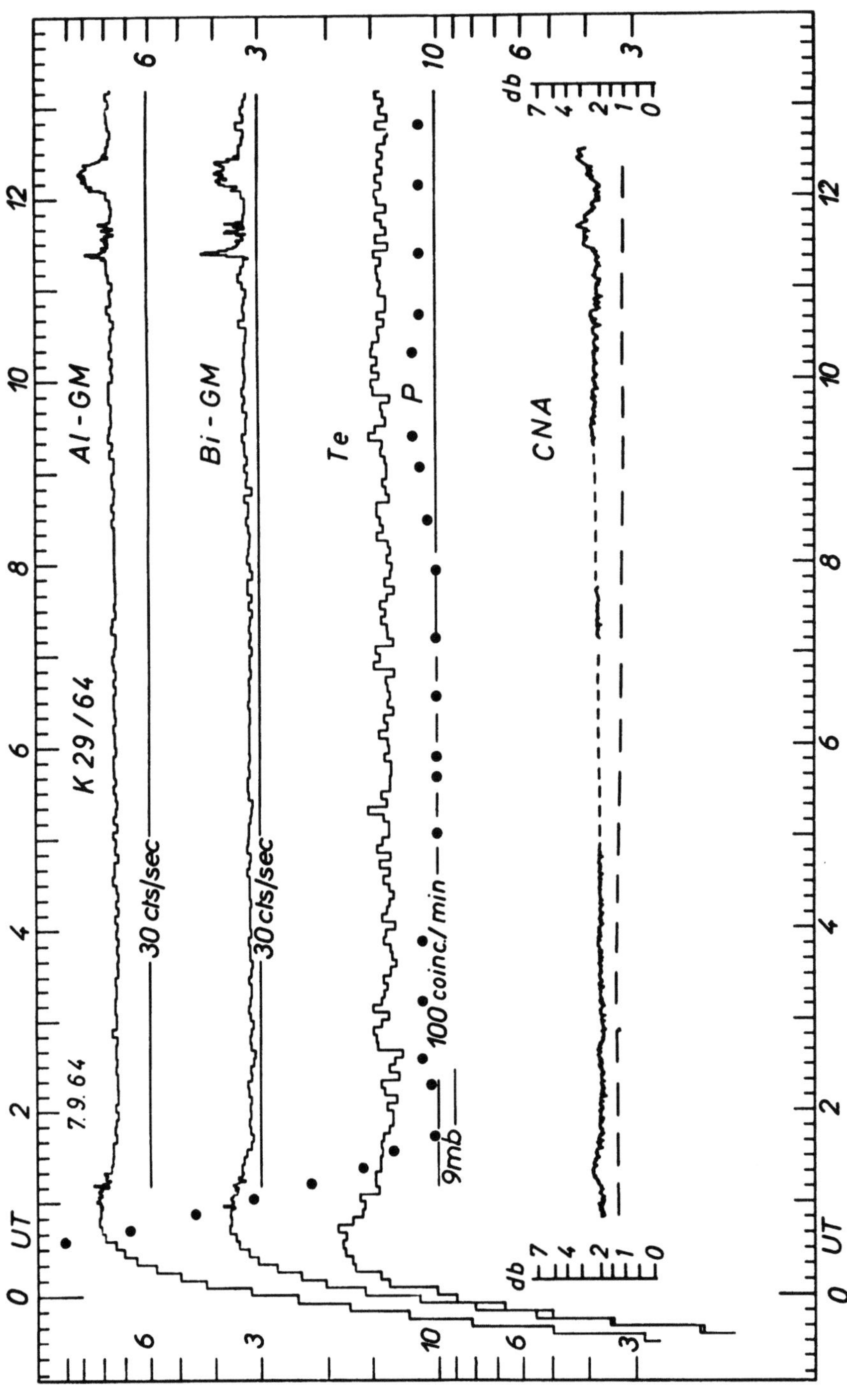

K 29/64
7.9.64
AI-GM
30 cts/sec
Bi-GM
30 cts/sec
Te
P
100 coinc/min
9mb
CNA
db
UT

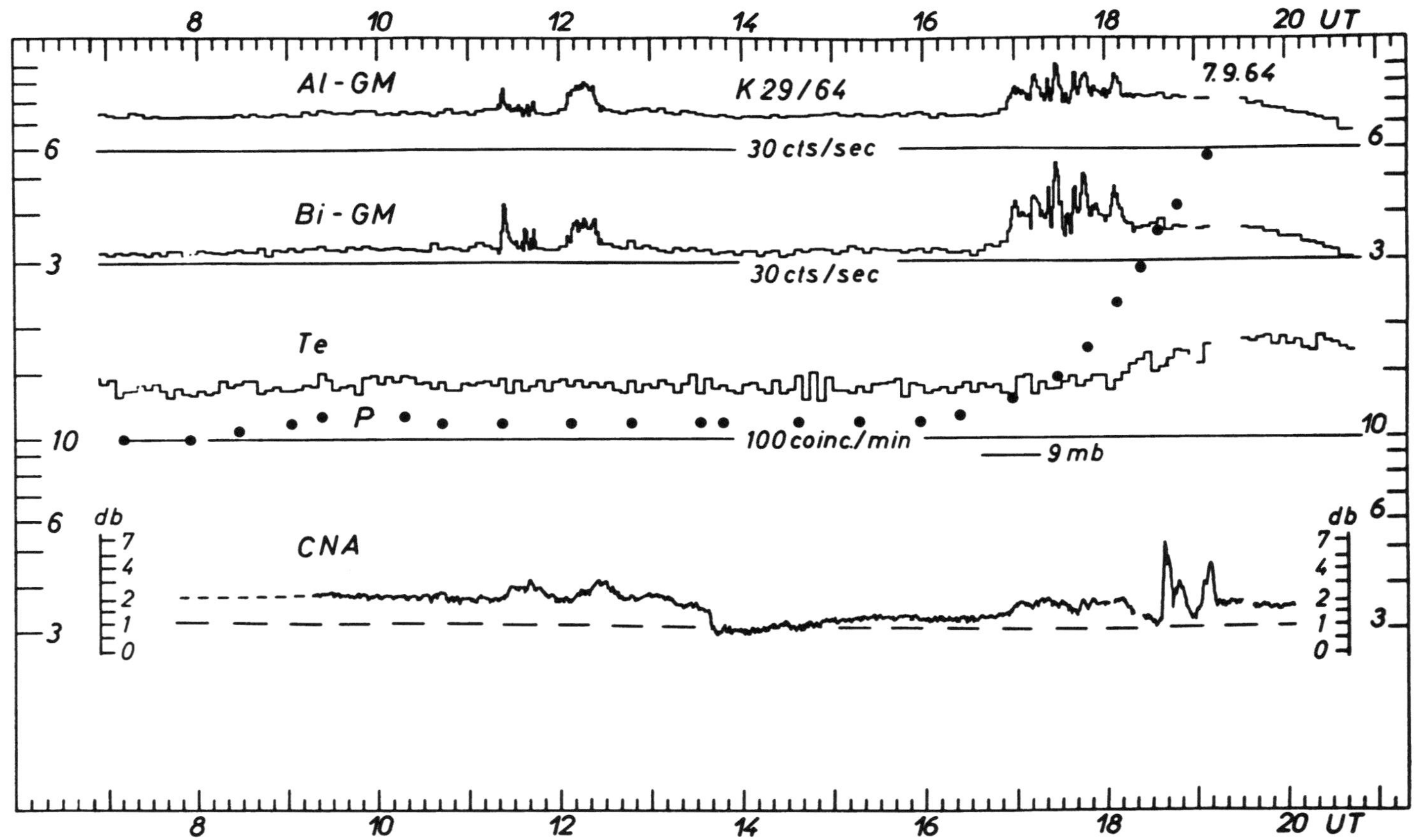

8 10 12 14 16 18 20 UT
Al - GM
K 29 / 64
7.9.64
6
30 cts/sec
6
Bi - GM
3
30 cts/sec
3
Te
P
100 coinc./min
9 mb
10
10
6
db
7
4
2
1
0
CNA
db 6
7
4
2
1
0
3
3
8 10 12 14 16 18 20 UT

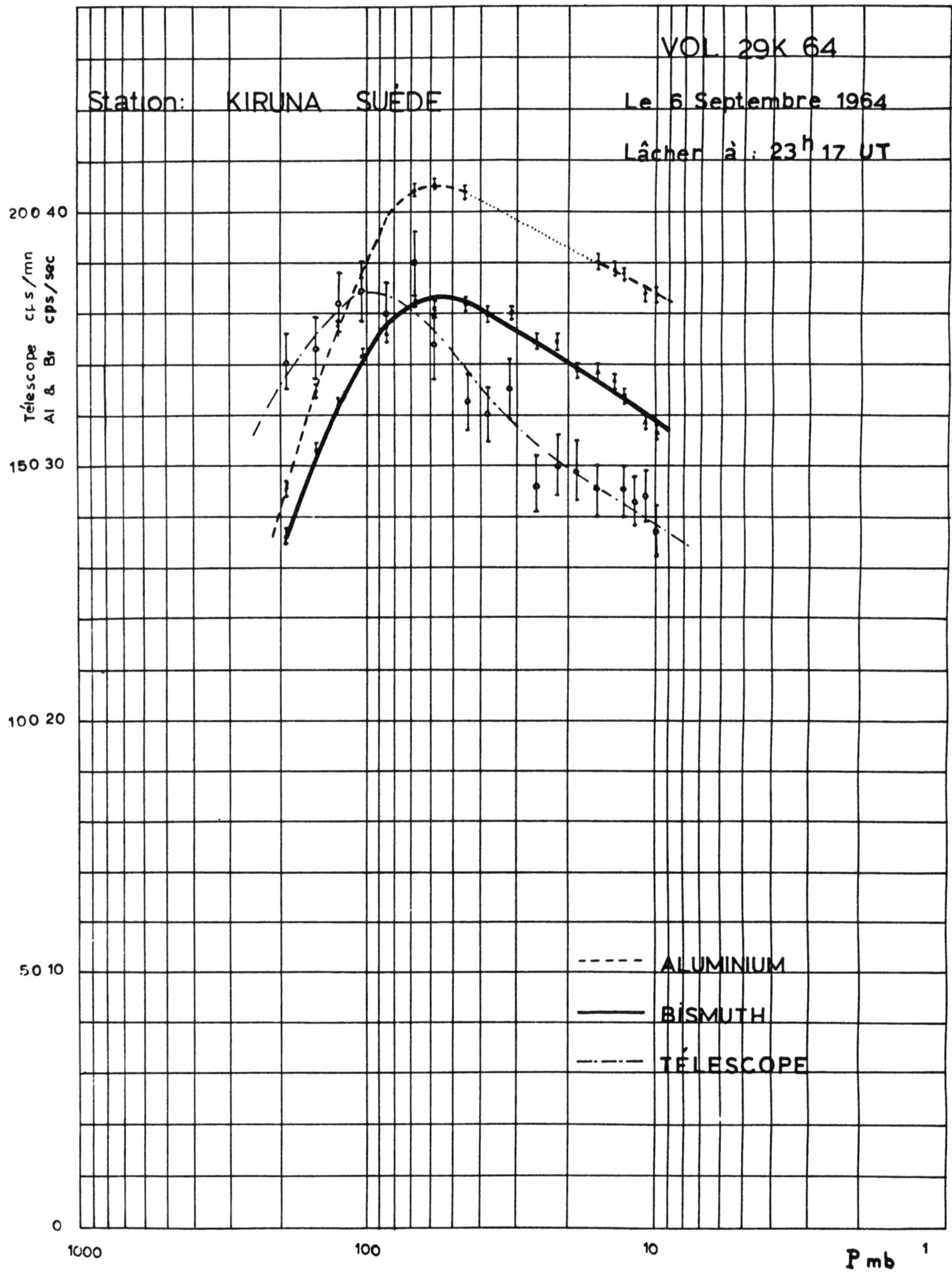

VOL 29K 64
Station: KIRUNA SUÈDE
Le 6 Septembre 1964
Lâcher à : 23 h 17 UT
Télescope cps/mn
Al & Bi cps/sec
200 40
150 30
100 20
50 10
0
1000
100
10
P mb
1
ALUMINIUM
BISMUTH
TÉLESCOPE

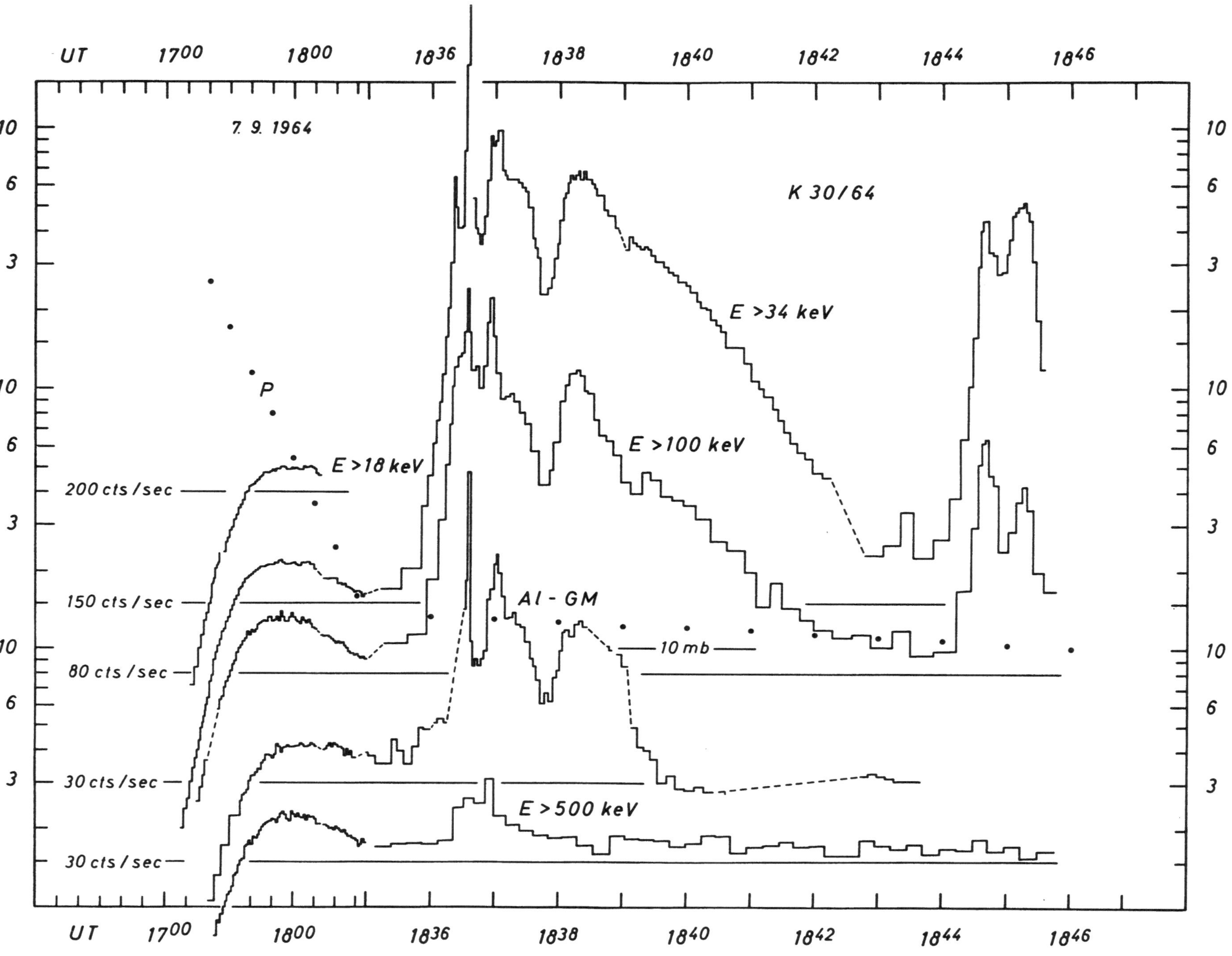

UT
7.9.1964
K 30/64
E >34 keV
E >100 keV
E >18 keV
200 cts/sec
150 cts/sec
80 cts/sec
30 cts/sec
30 cts/sec
Al - GM
10 mb
E >500 keV
P

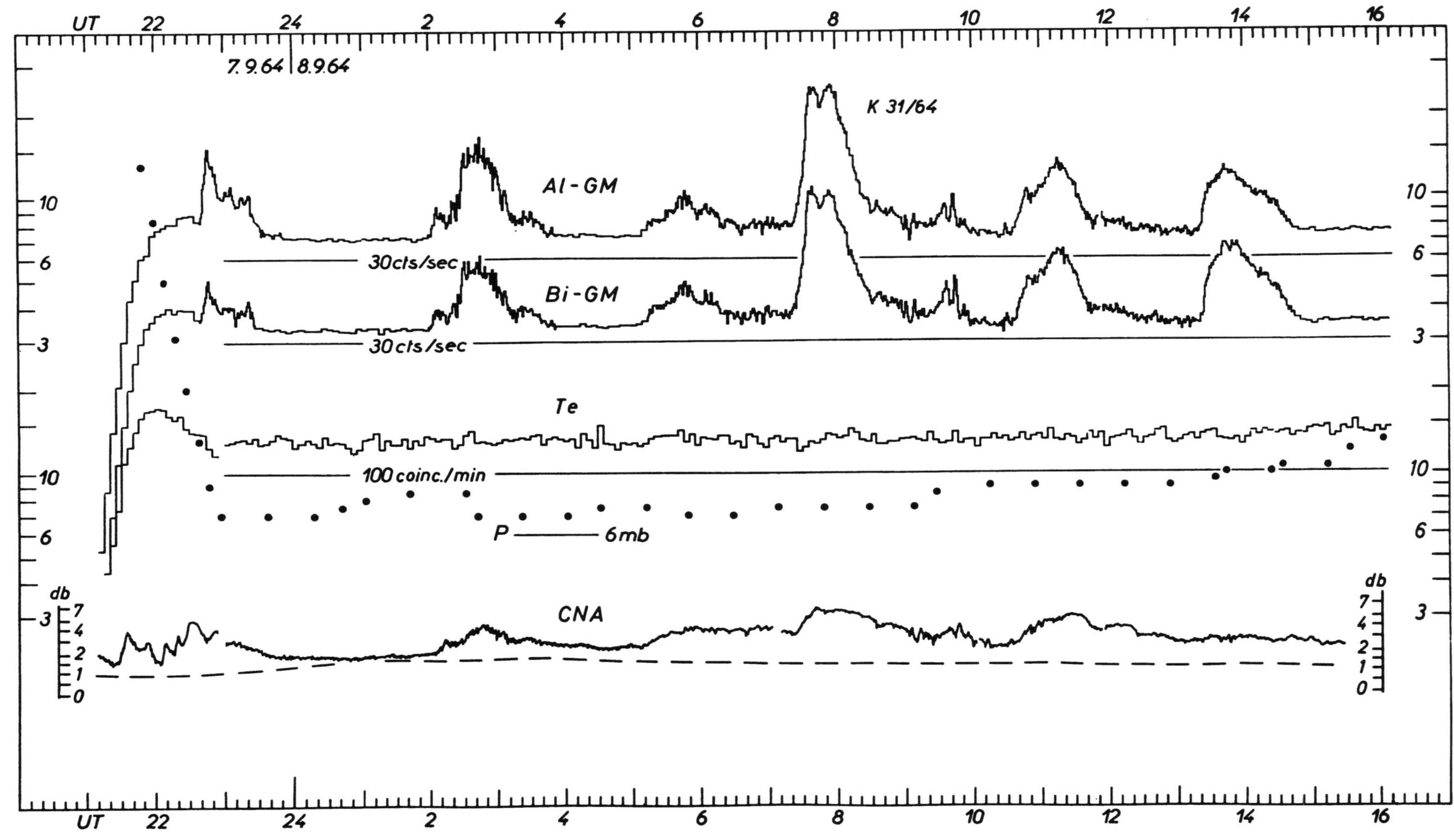

UT
7.9.64 | 8.9.64
K 31/64
Al-GM
30 cts/sec
Bi-GM
30 cts/sec
Te
100 coinc./min
P —— 6mb
db
CNA

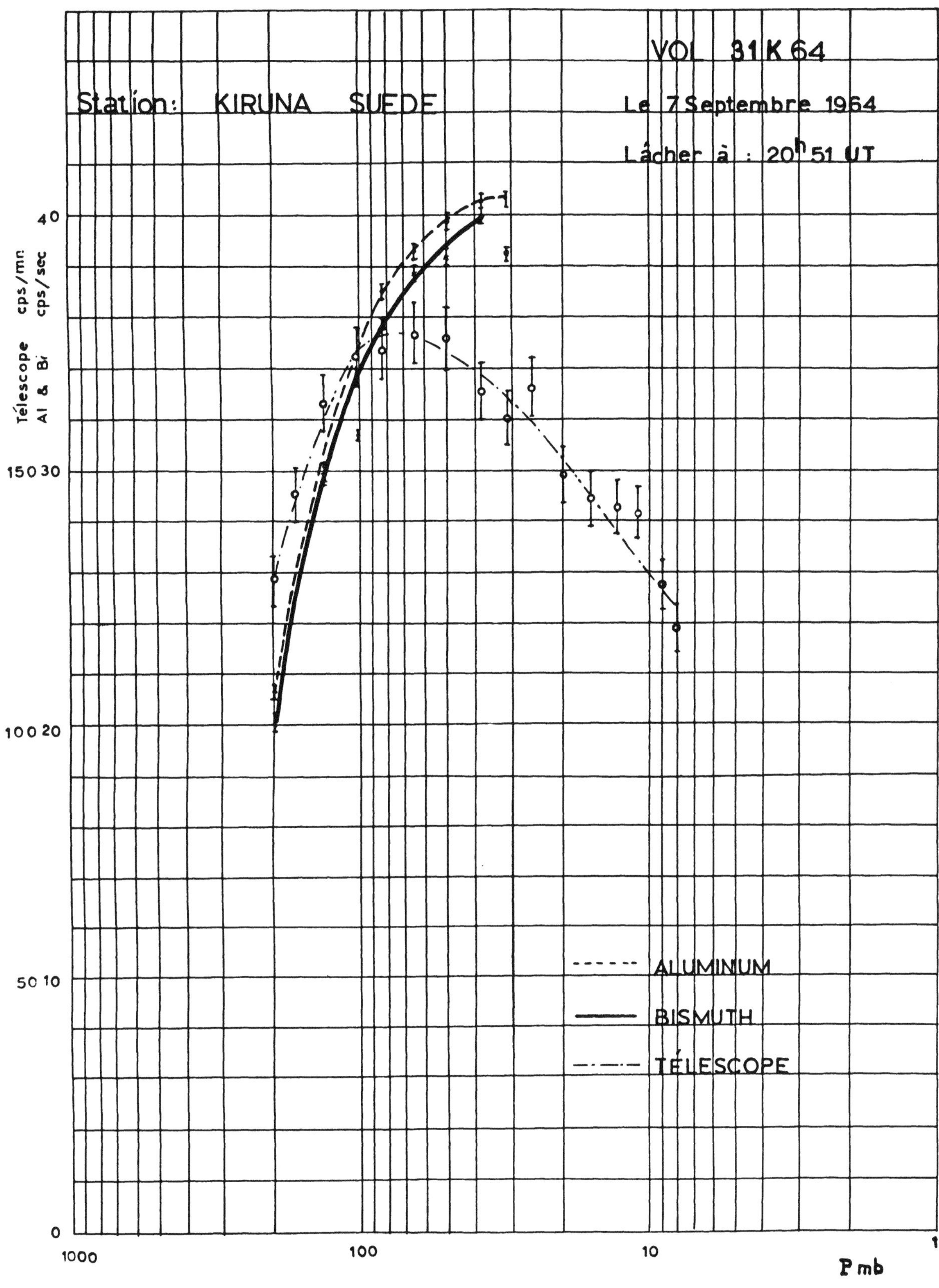

VOL 31 K 64
Station: KIRUNA SUEDE
Le 7 Septembre 1964
Lâcher à : 20h 51 UT
Télescope cps/mn
Al & Bi cps/sec
40
30
20
10
150
100
50
0
1000
100
10
1
P mb
ALUMINUM
BISMUTH
TÉLESCOPE

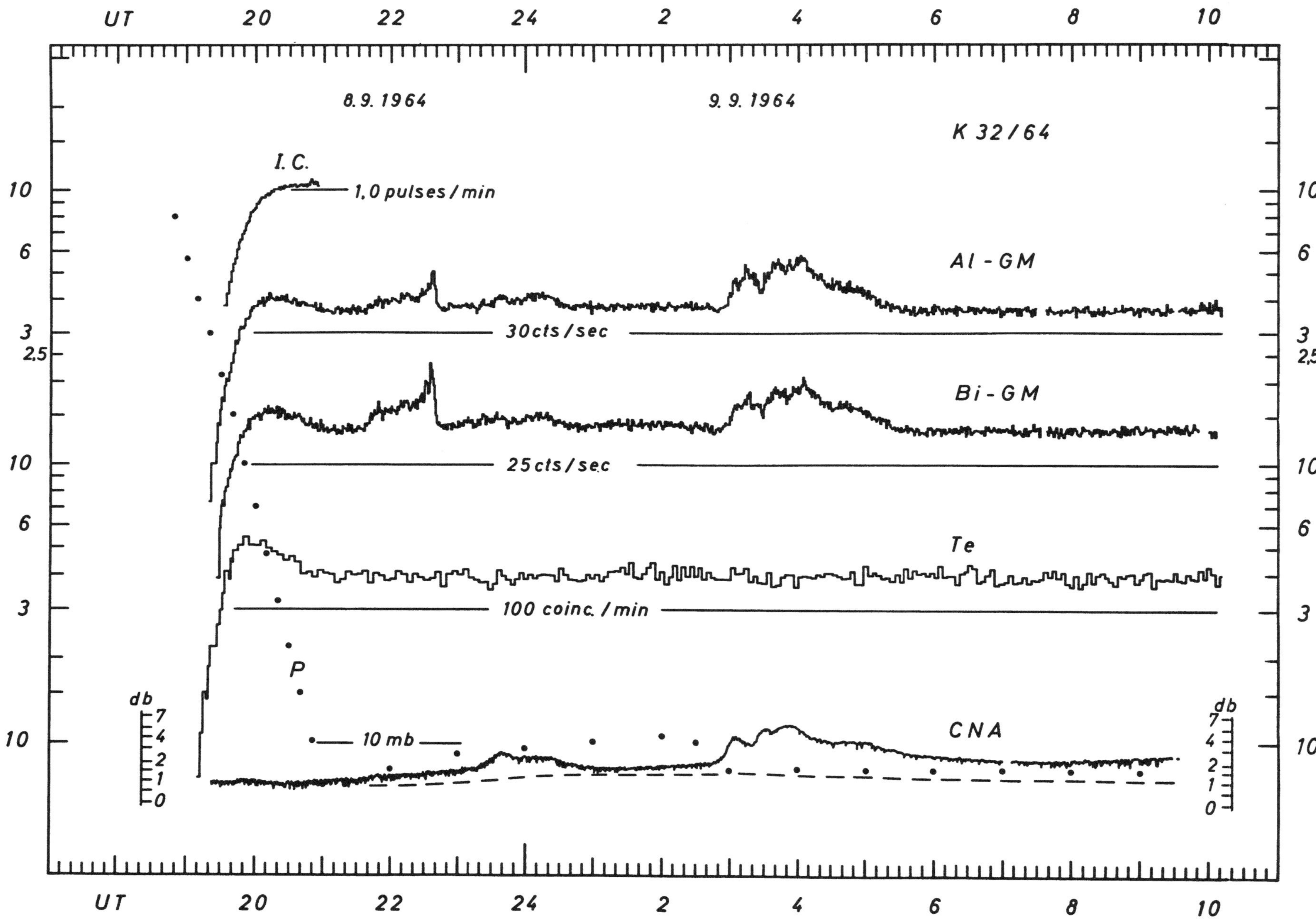
UT 20 22 24 2 4 6 8 10
8.9.1964
9.9.1964
K 32/64
I.C.
1.0 pulses/min
AI-GM
30 cts/sec
Bi-GM
25 cts/sec
Te
100 coinc./min
P
db 7 4 2 1 0
10 mb
CNA
db 7 4 2 1 0
UT 20 22 24 2 4 6 8 10

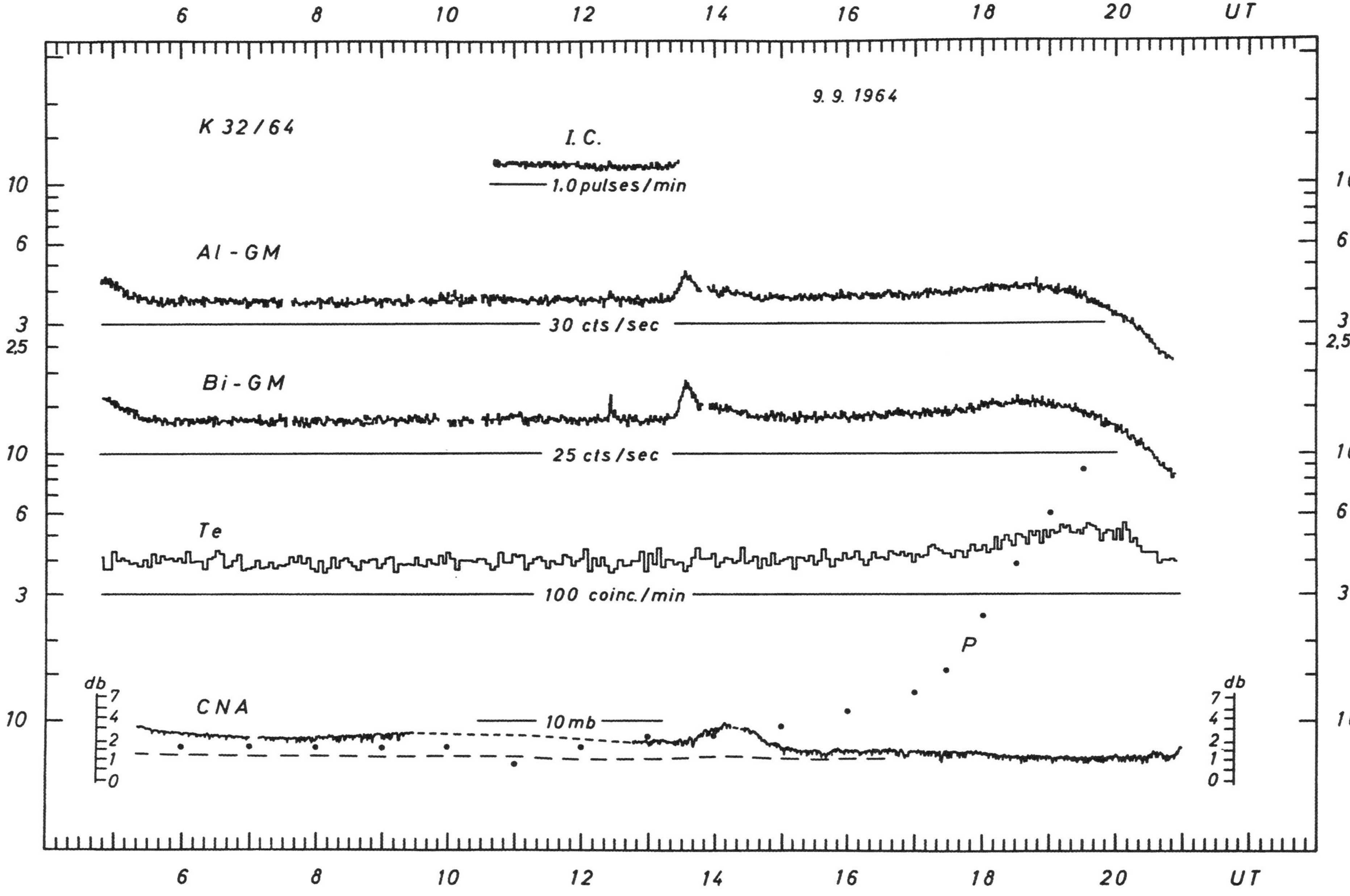

6   8   10   12   14   16   18   20   UT
9. 9. 1964
K 32/64
I.C.
1.0 pulses/min
Al - GM
30 cts/sec
Bi - GM
25 cts/sec
Te
100 coinc./min
db 7 4 2 1 0
CNA
10mb
P
10 6 3 2,5 10 6 3 10

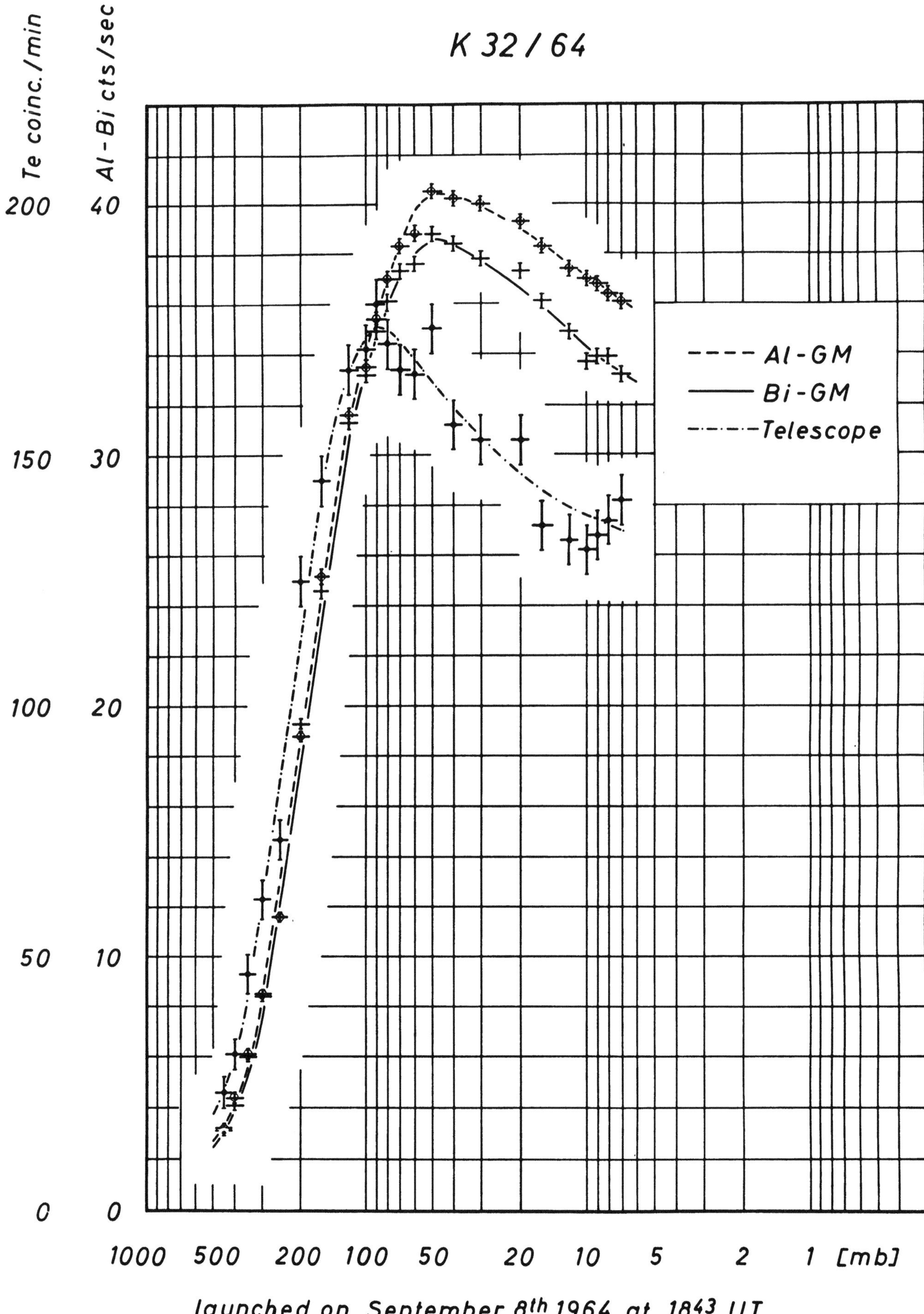

K 32 / 64
Te coinc./min
Al - Bi cts/sec
200
150
100
50
0
40
30
20
10
0
Al - GM
Bi - GM
Telescope
1000
500
200
100
50
20
10
5
2
1 [mb]
launched on September 8th 1964 at 1843 UT

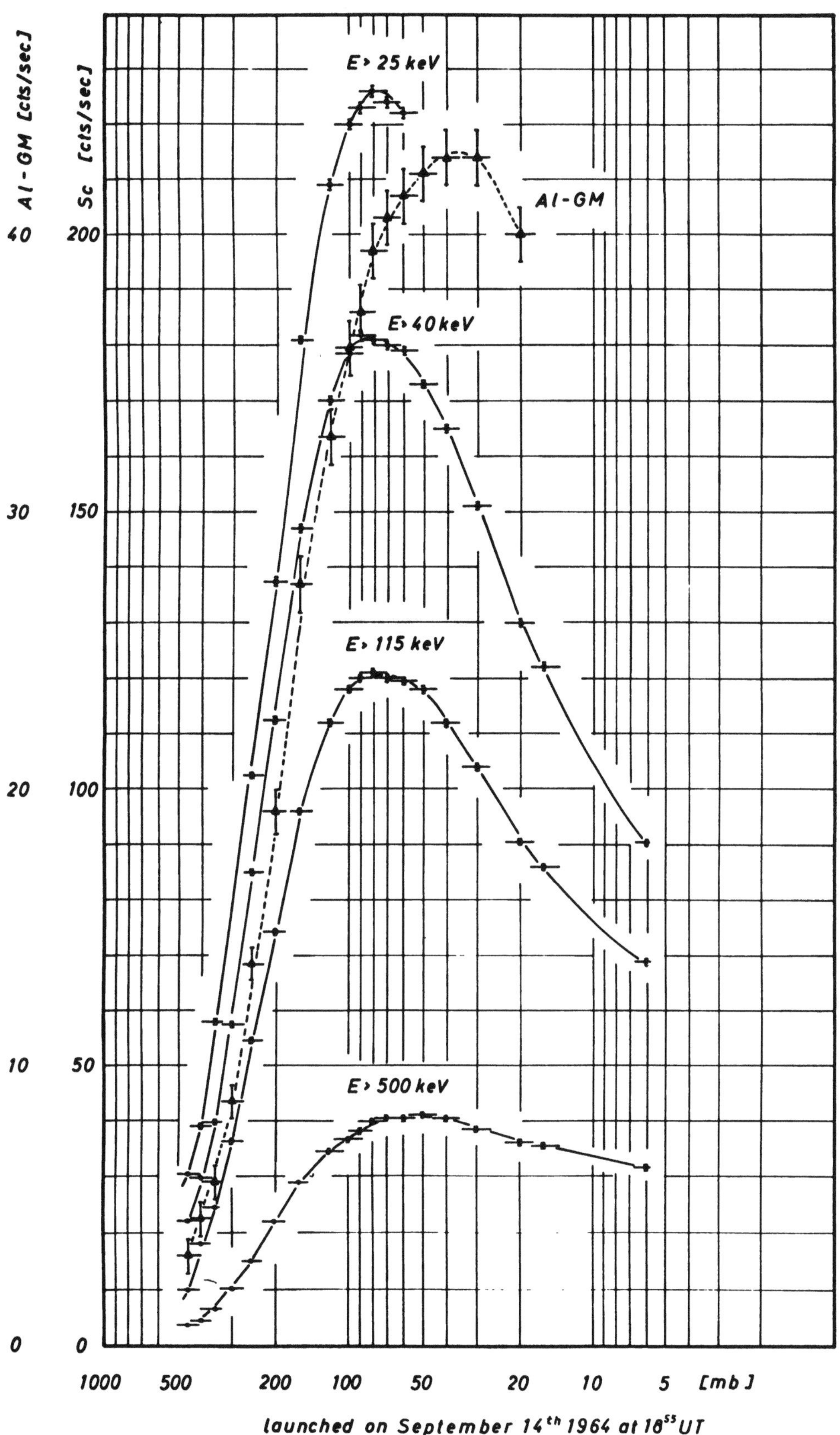
K 33 / 64
Al - GM [cts/sec]
Sc [cts/sec]
E > 25 keV
Al - GM
E > 40 keV
E > 115 keV
E > 500 keV
1000 500 200 100 50 20 10 5 [mb]
launched on September 14th 1964 at 18⁵⁵ UT
40
30
20
10
0
200
150
100
50
0

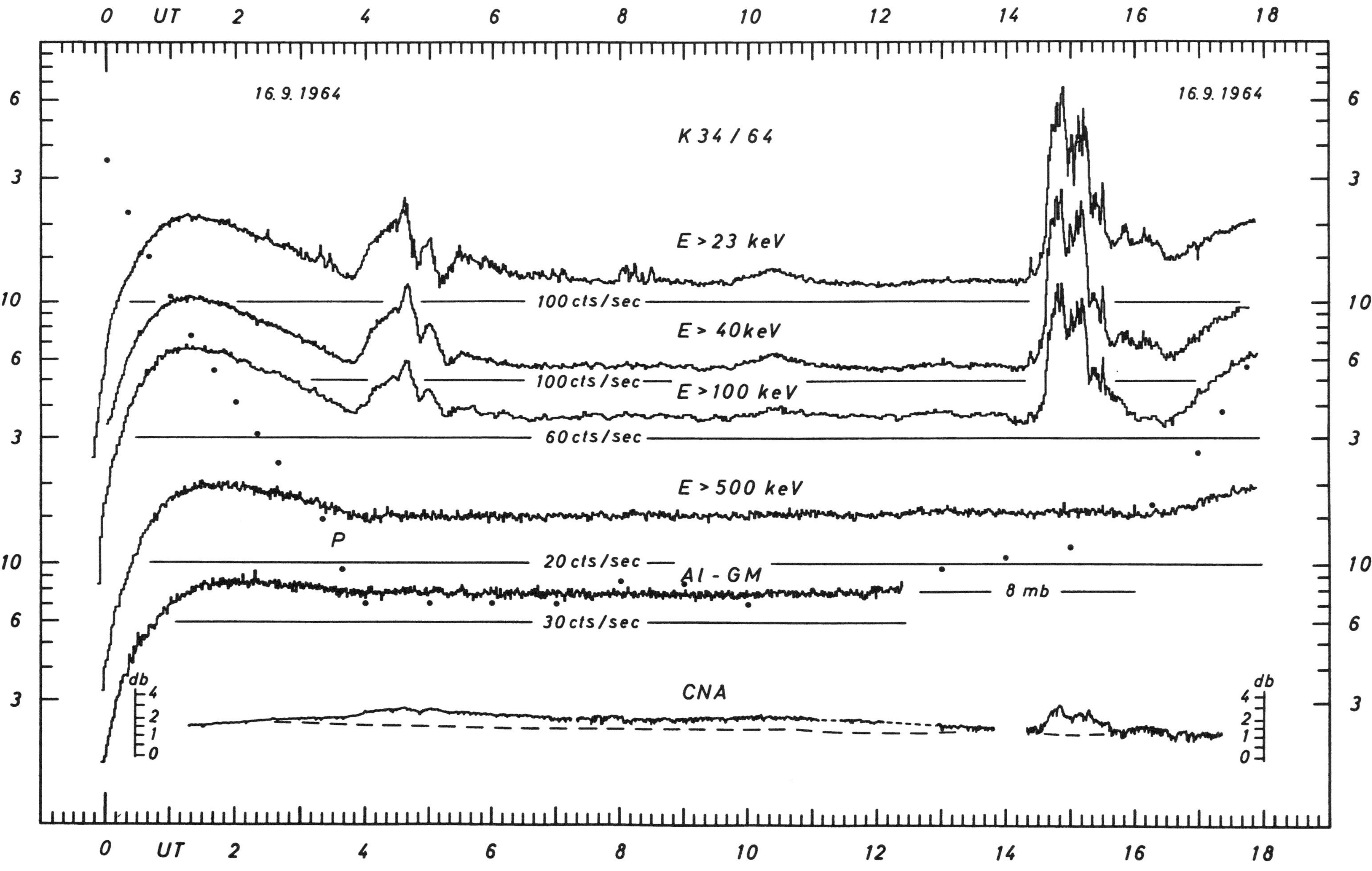

0 UT 2 4 6 8 10 12 14 16 18
16. 9. 1964
16. 9. 1964
K 34 / 64
E > 23 keV
100 cts/sec
E > 40 keV
100 cts/sec
E > 100 keV
60 cts/sec
E > 500 keV
P
20 cts/sec
Al - GM
8 mb
30 cts/sec
CNA
db
4 2 1 0
0 UT 2 4 6 8 10 12 14 16 18

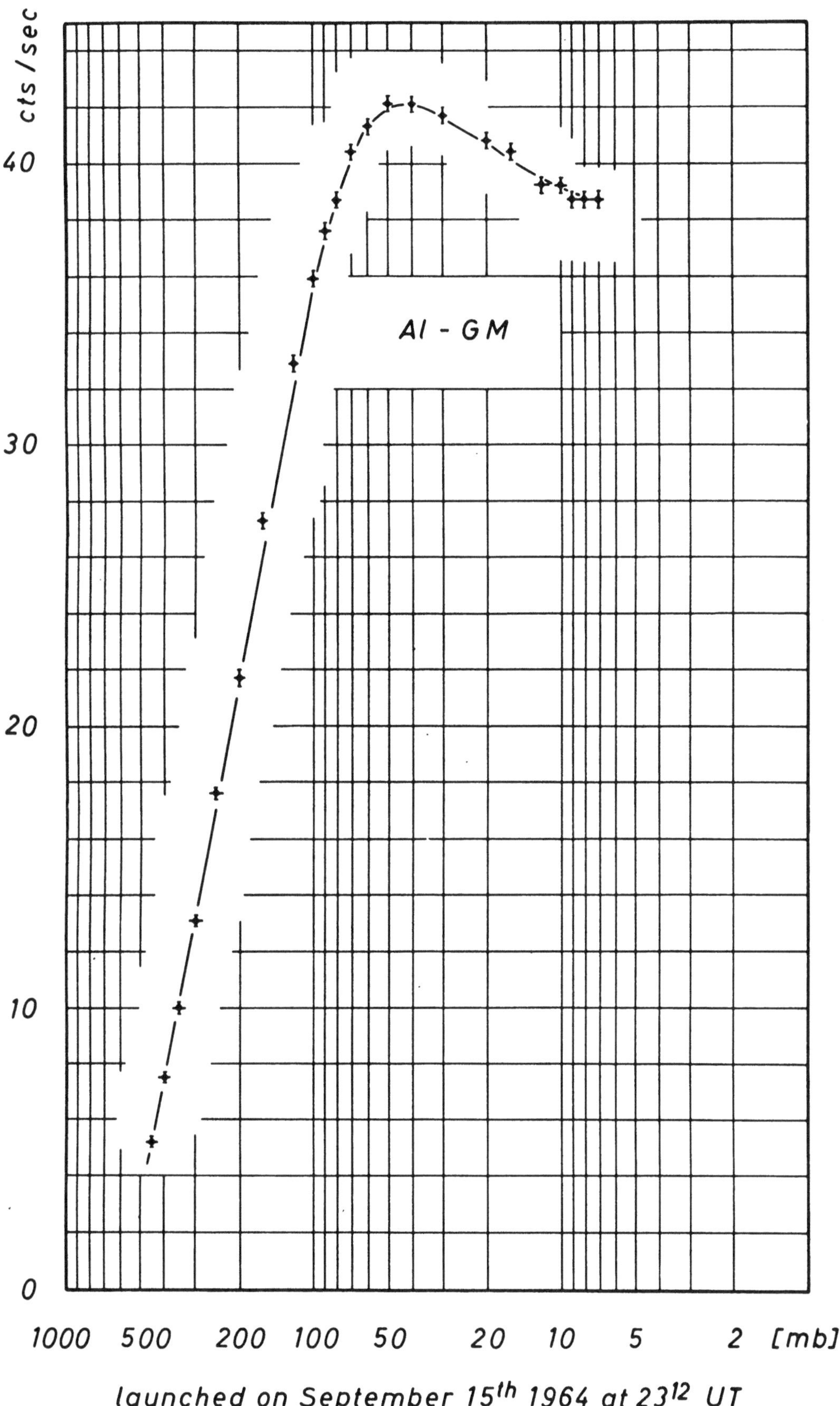

K 34 / 64
cts / sec
40
30
20
10
0
Al - GM
1000 500 200 100 50 20 10 5 2 [mb]
launched on September 15th 1964 at 23^12 UT

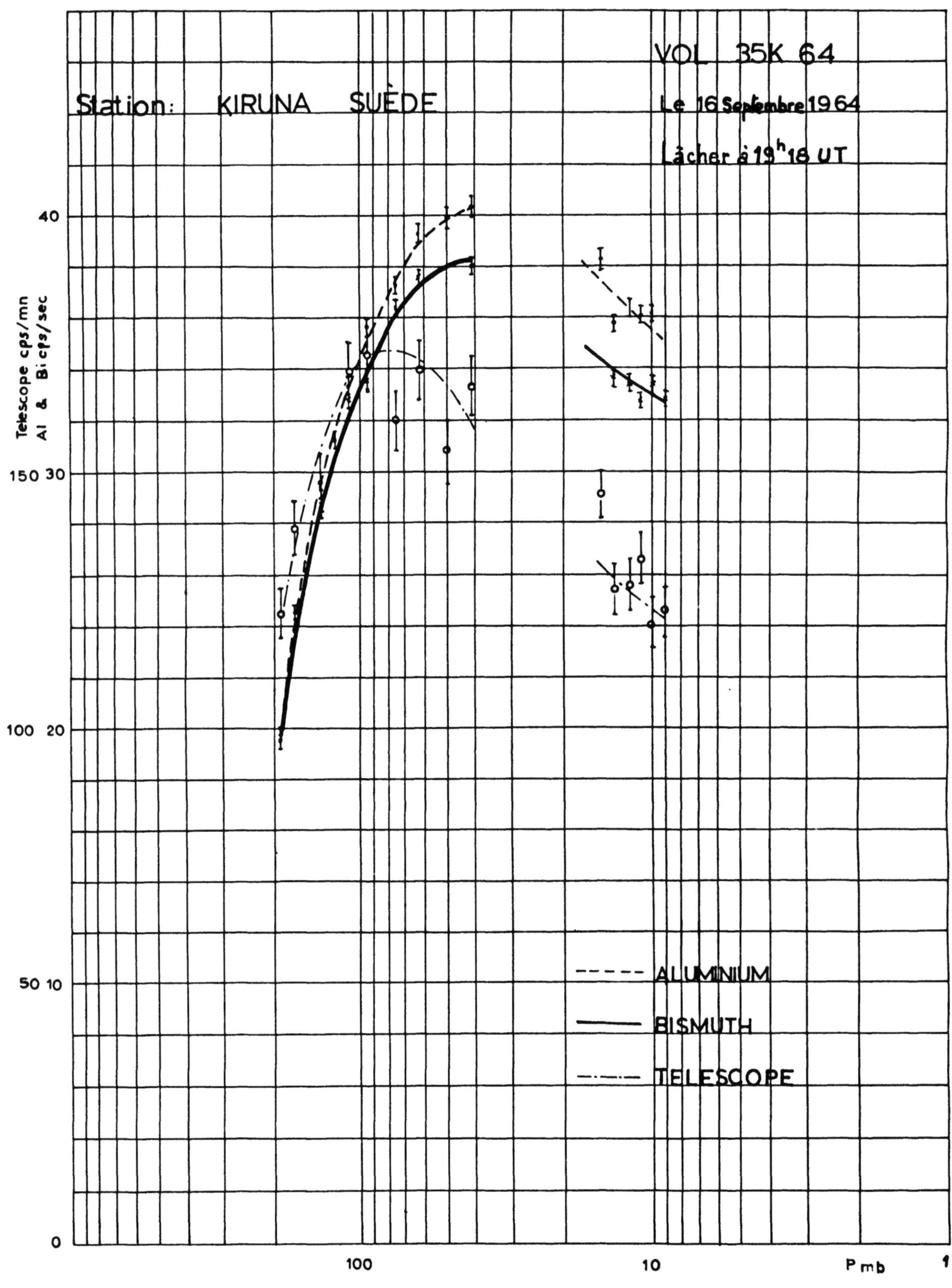
VOL 35K 64
Station: KIRUNA SUÈDE
Le 16 Septembre 1964
Lâcher à 19h18 UT
Telescope cps/mn
Al & Bi cps/sec
40
150 30
100 20
50 10
0
100
10
P mb
----- ALUMINIUM
——— BISMUTH
—·—·— TELESCOPE

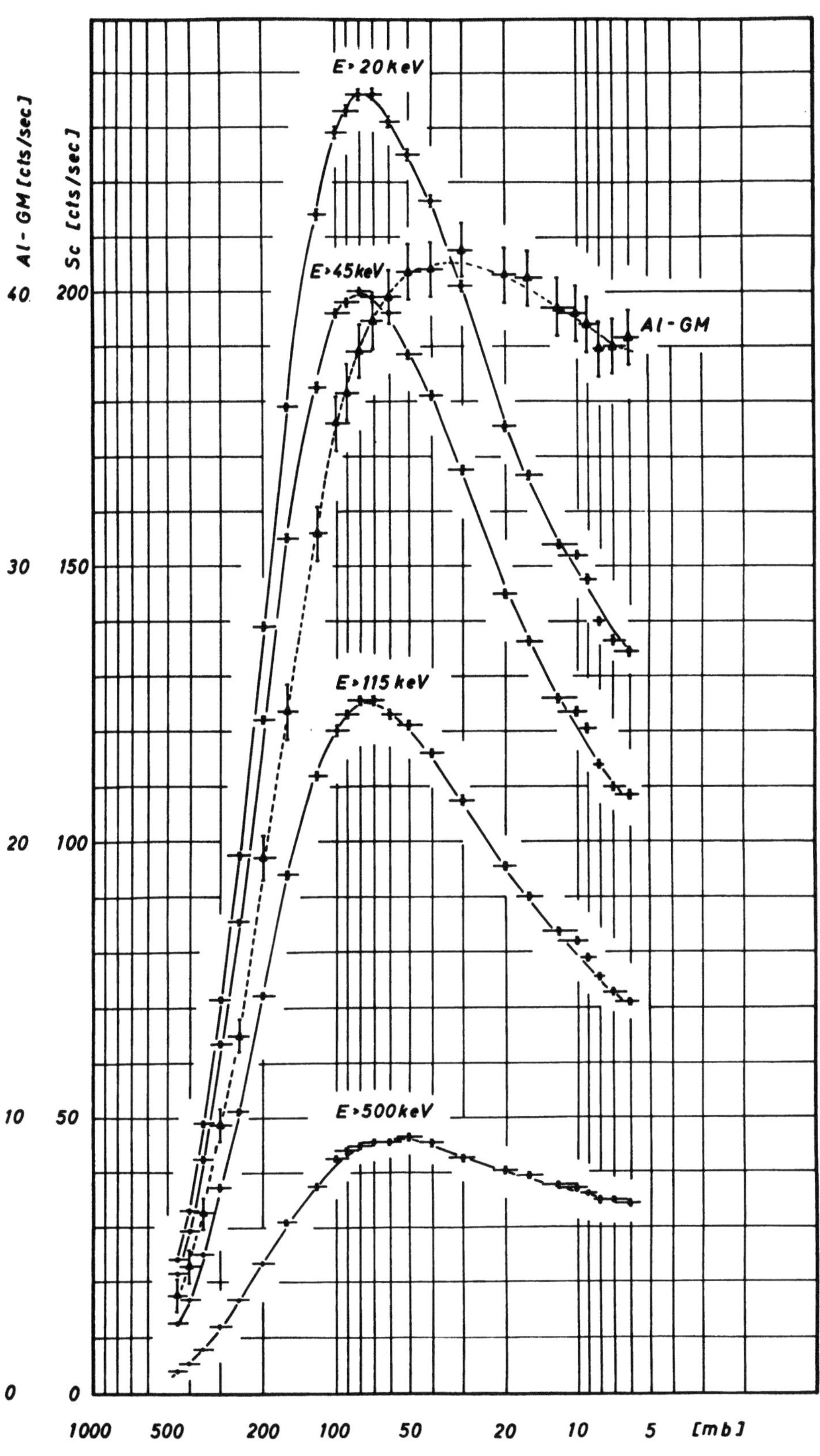

K 36 / 64
Al - GM [cts/sec]
Sc [cts/sec]
E>20keV
E>45keV
E>115 keV
E>500keV
Al-GM
40 200
30 150
20 100
10 50
0 0
1000 500 200 100 50 20 10 5 [mb]
launched on September 18th 1964 at 22^12 UT

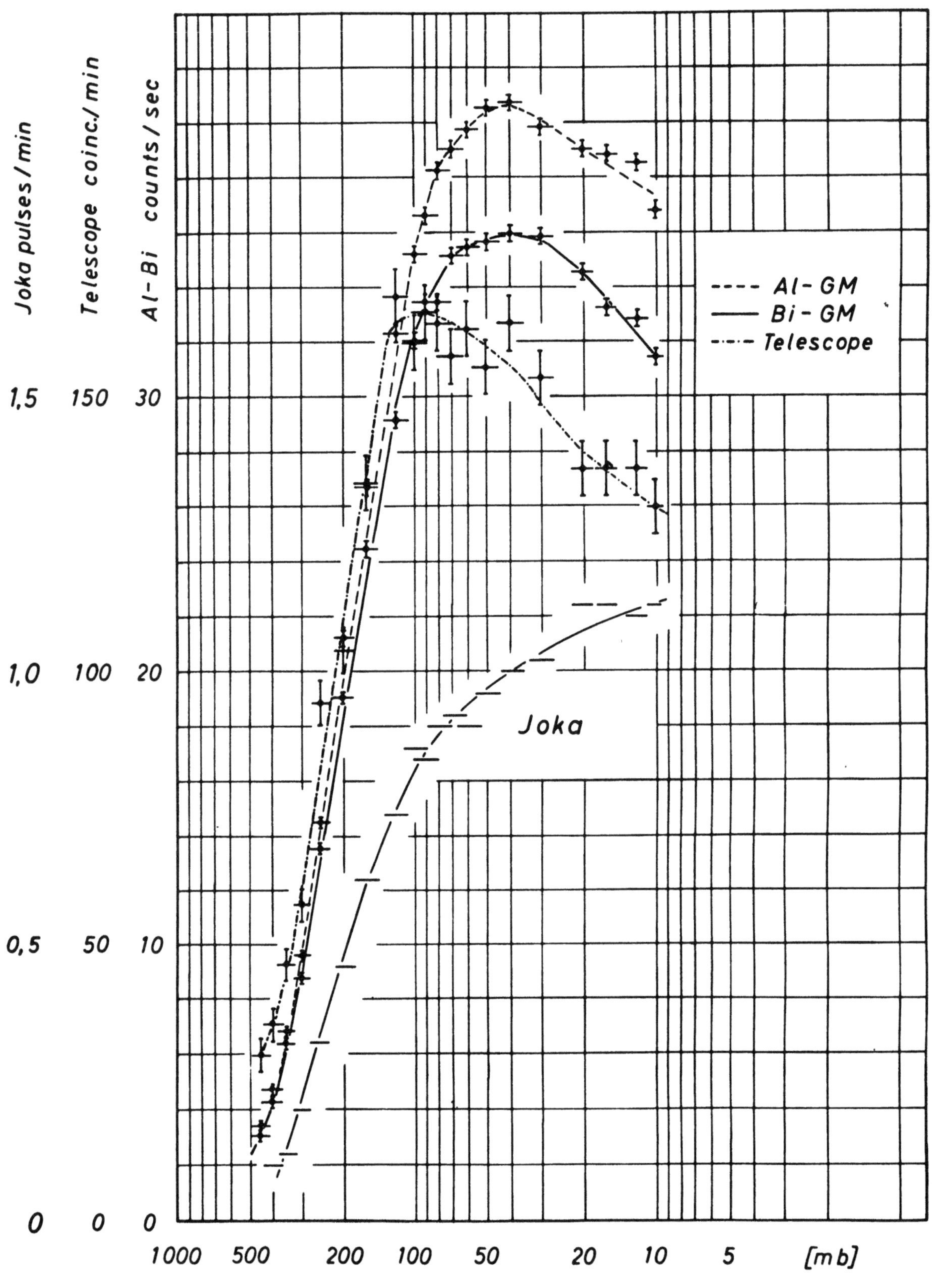

K 37 / 64
Joka pulses / min
Telescope coinc./ min
Al – Bi counts / sec
Al - GM
Bi - GM
Telescope
Joka
1,5
1,0
0,5
0
150
100
50
0
30
20
10
0
1000
500
200
100
50
20
10
5
[mb]
launched on September 22nd 1964 at 9⁴⁷ UT

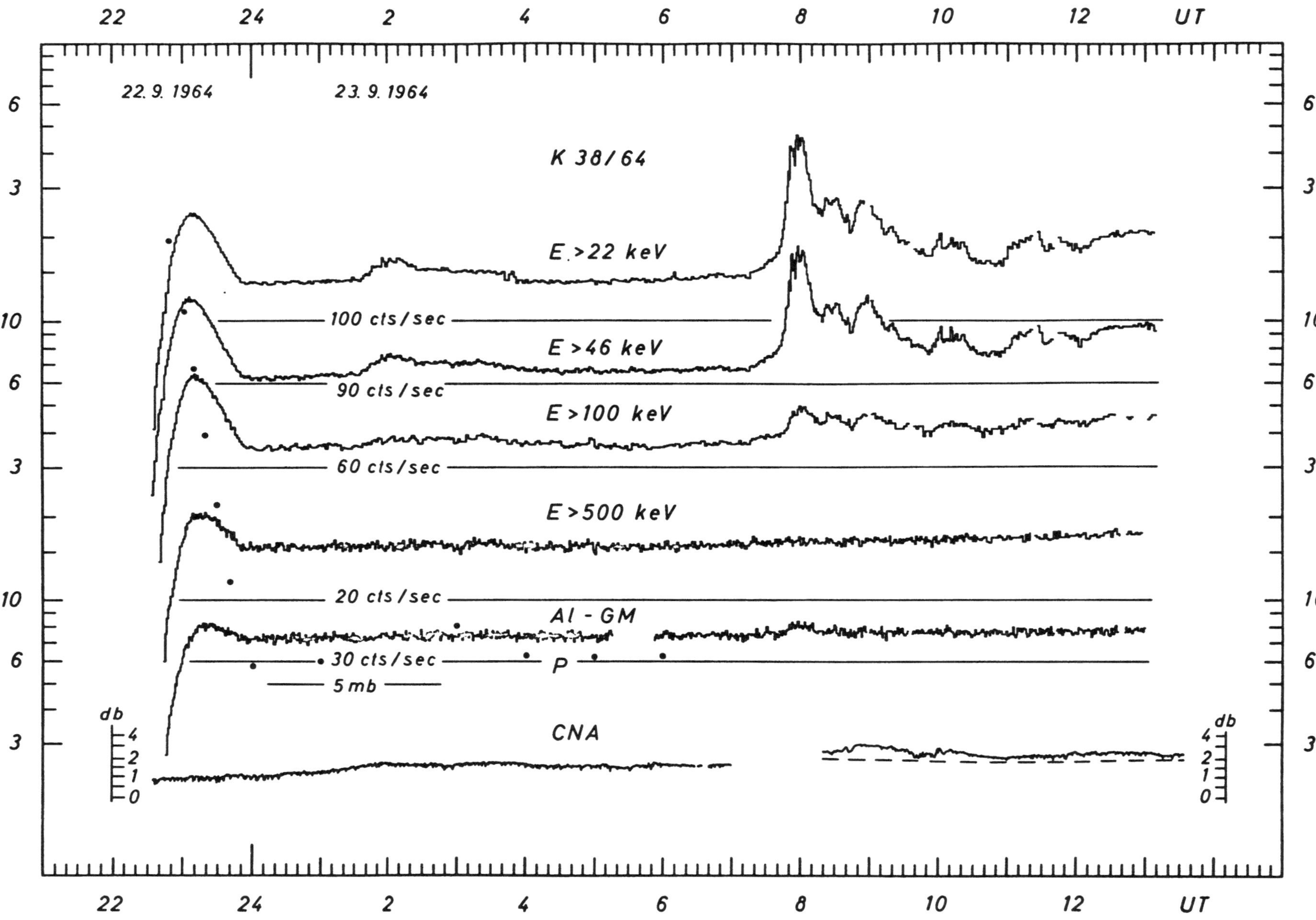

22 24 2 4 6 8 10 12 UT
22. 9. 1964
23. 9. 1964
K 38/64
E >22 keV
100 cts/sec
E >46 keV
90 cts/sec
E >100 keV
60 cts/sec
E >500 keV
20 cts/sec
Al - GM
30 cts/sec
P
5 mb
db
4
2
1
0
CNA
db
4
2
1
0
6
3
10
6
3
10
6
3
22 24 2 4 6 8 10 12 UT

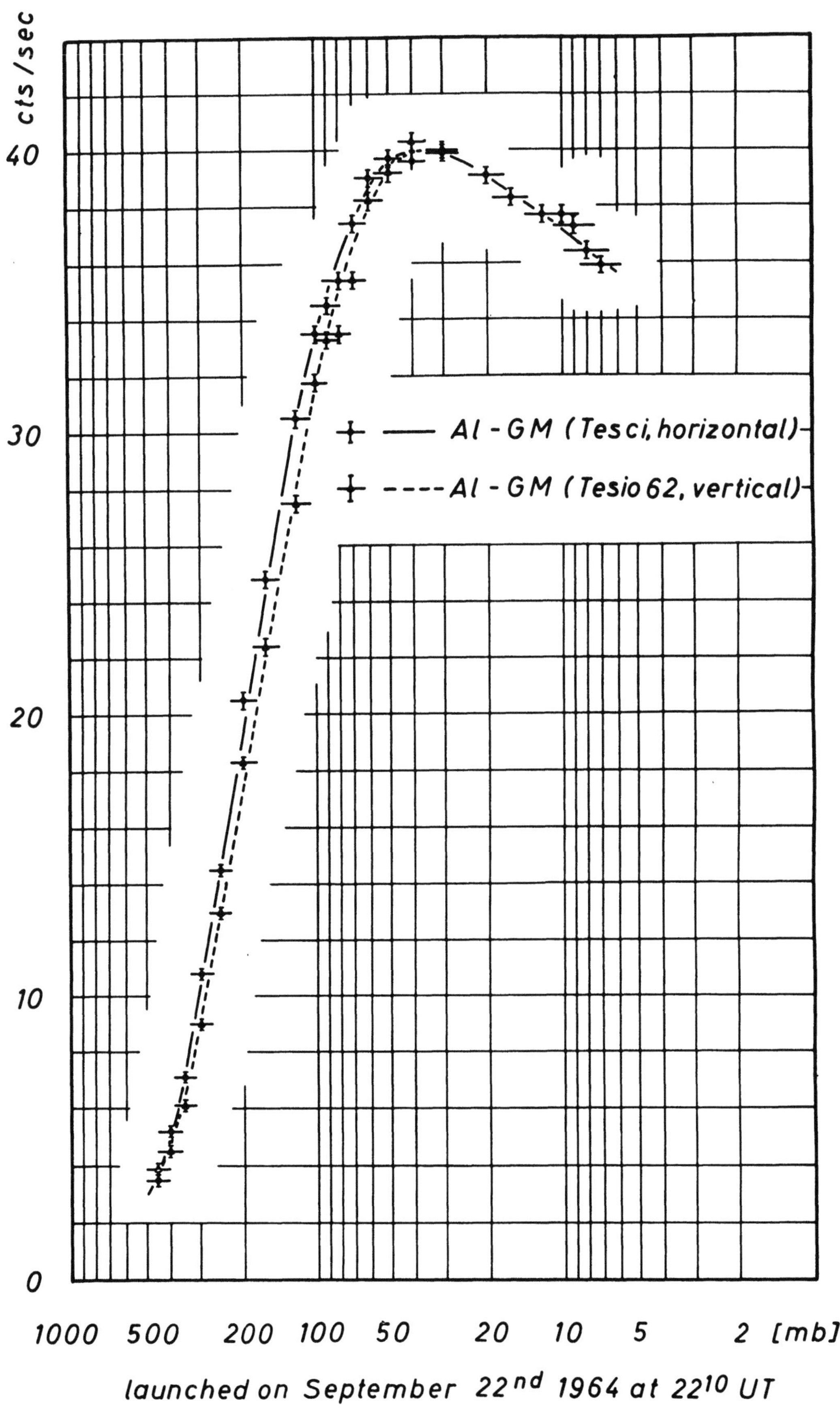

K 38 / 64
cts/sec
40
30
20
10
0
Al - GM (Tesci, horizontal)
Al - GM (Tesio 62, vertical)
1000 500 200 100 50 20 10 5 2 [mb]
launched on September 22nd 1964 at 22:10 UT

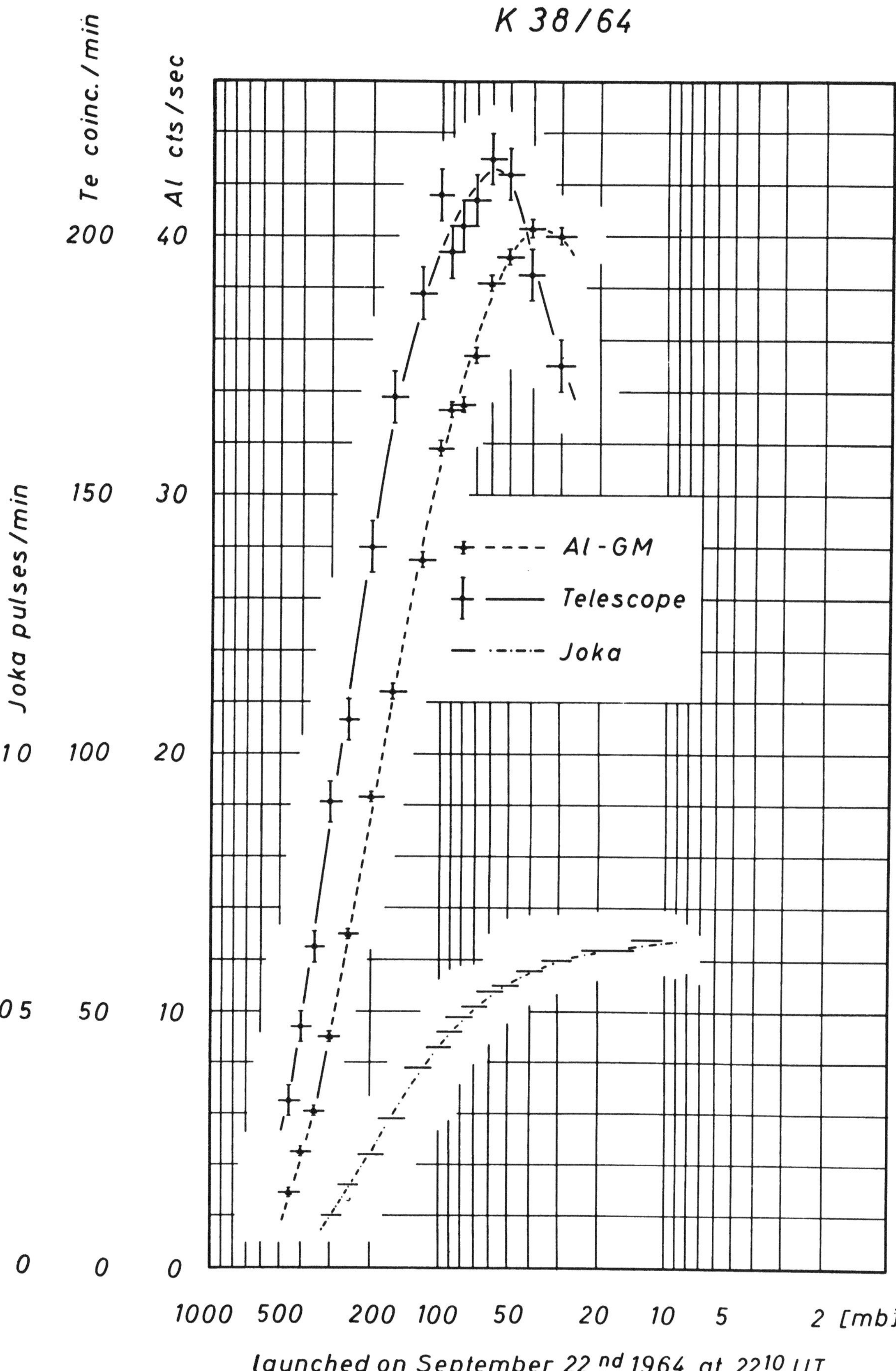
K 38/64
Te coinc./min
Al cts/sec
Joka pulses/min
Al-GM
Telescope
Joka
1000 500 200 100 50 20 10 5 2 [mb]
launched on September 22 nd 1964 at 22 10 UT

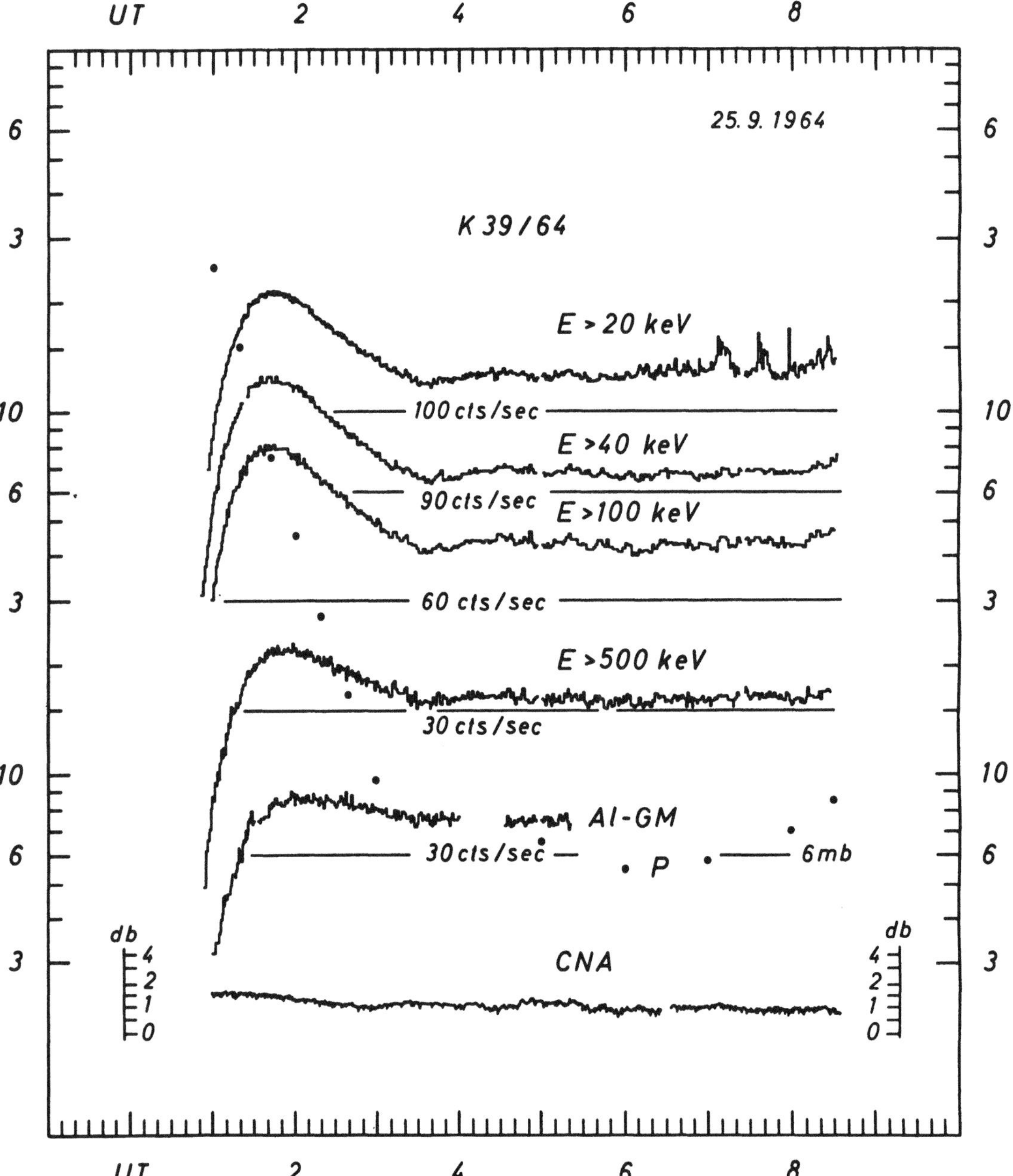

UT
2
4
6
8
25. 9. 1964
K 39 / 64
E > 20 keV
100 cts/sec
E > 40 keV
90 cts/sec
E > 100 keV
60 cts/sec
E > 500 keV
30 cts/sec
Al-GM
30 cts/sec
P
6 mb
db
4
2
1
0
CNA
db
4
2
1
0
UT
2
4
6
8

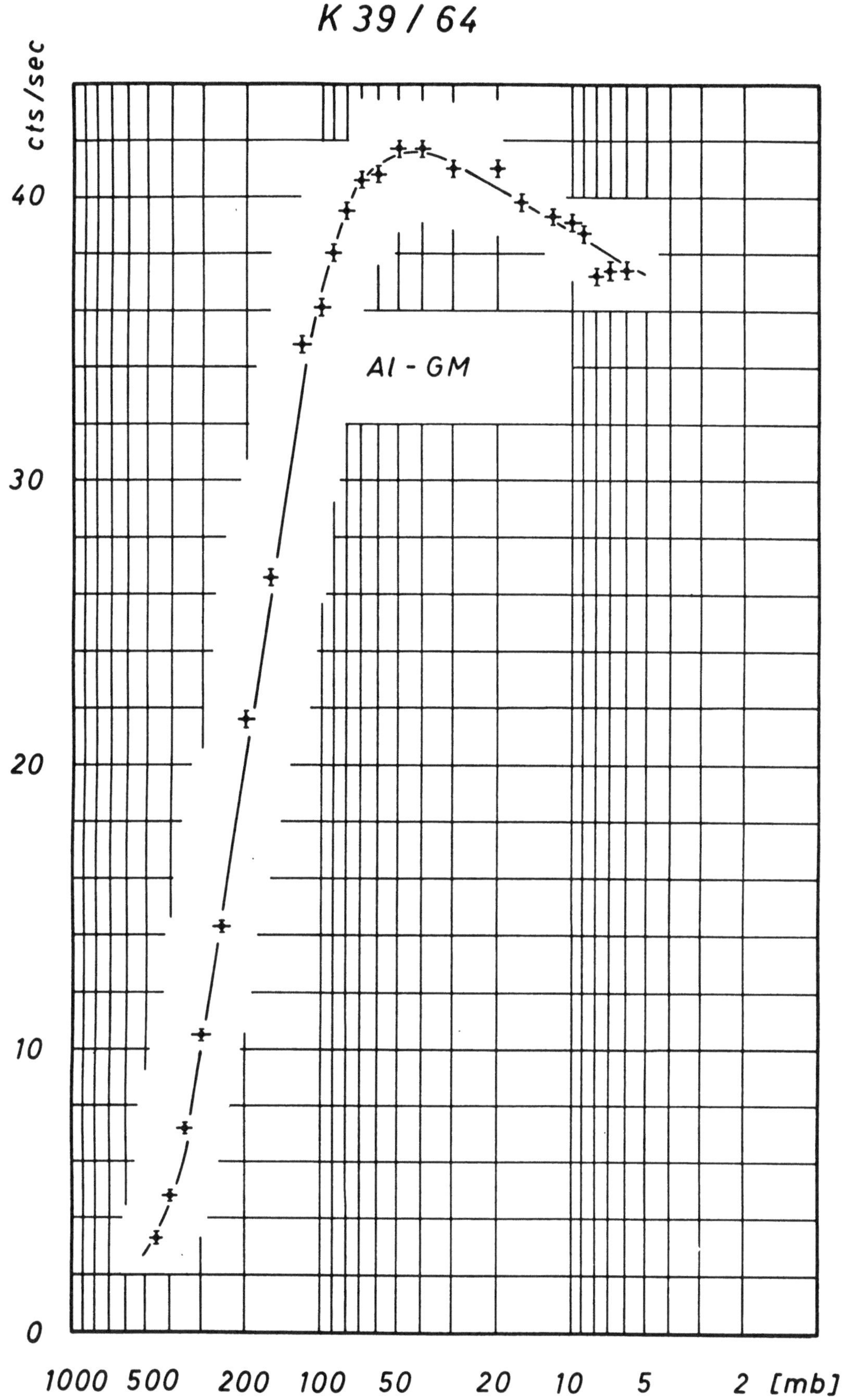

K 39 / 64
cts/sec
AI - GM
40
30
20
10
0
1000 500 200 100 50 20 10 5 2 [mb]
launched on September 25th 1964 at 0^10 11 T

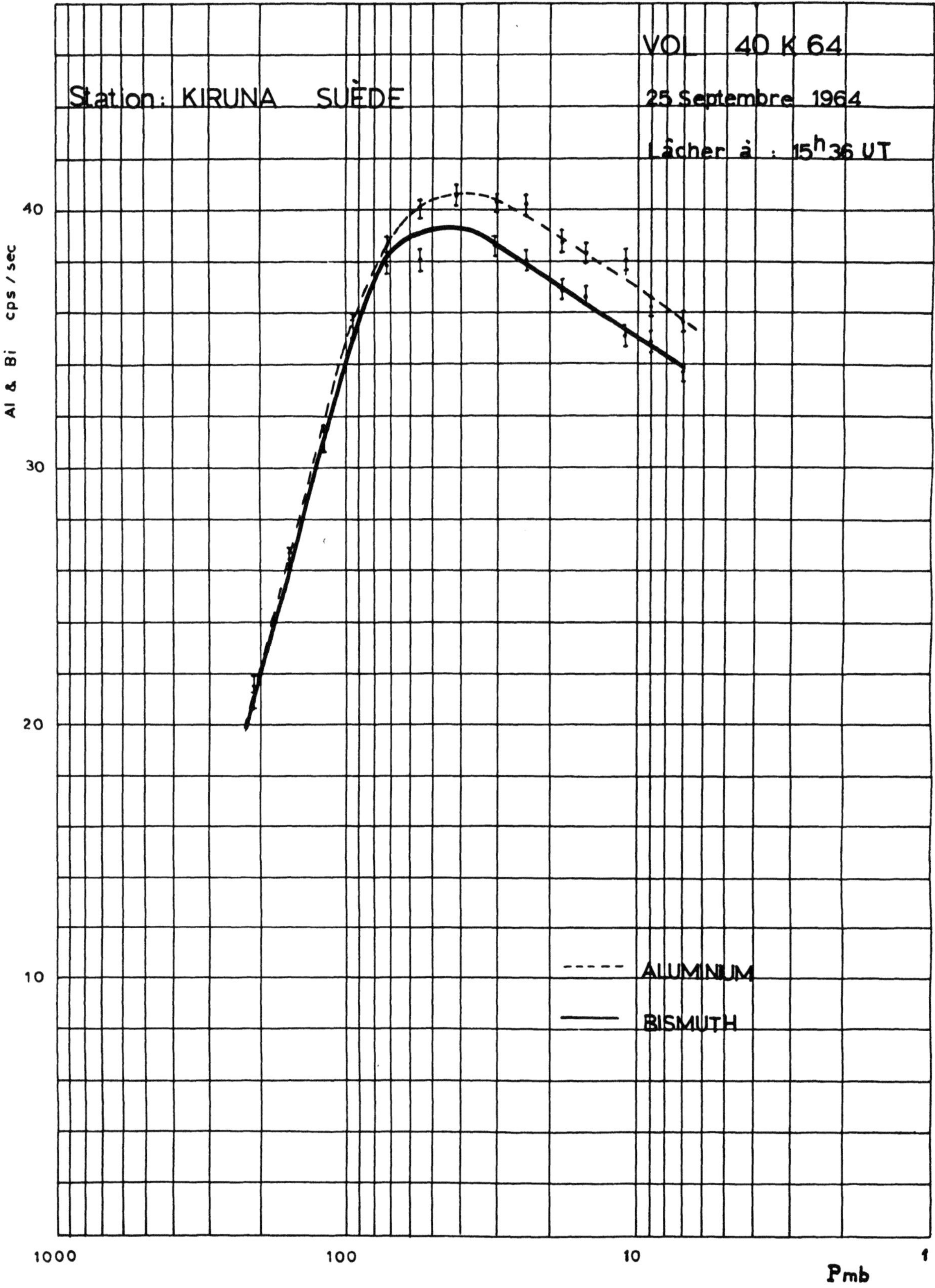

VOL 40 K 64
Station : KIRUNA SUÈDE
25 Septembre 1964
Lâcher à : 15h 36 UT
Al & Bi cps / sec
40
30
20
10
1000
100
10
1
Pmb
ALUMINIUM
BISMUTH

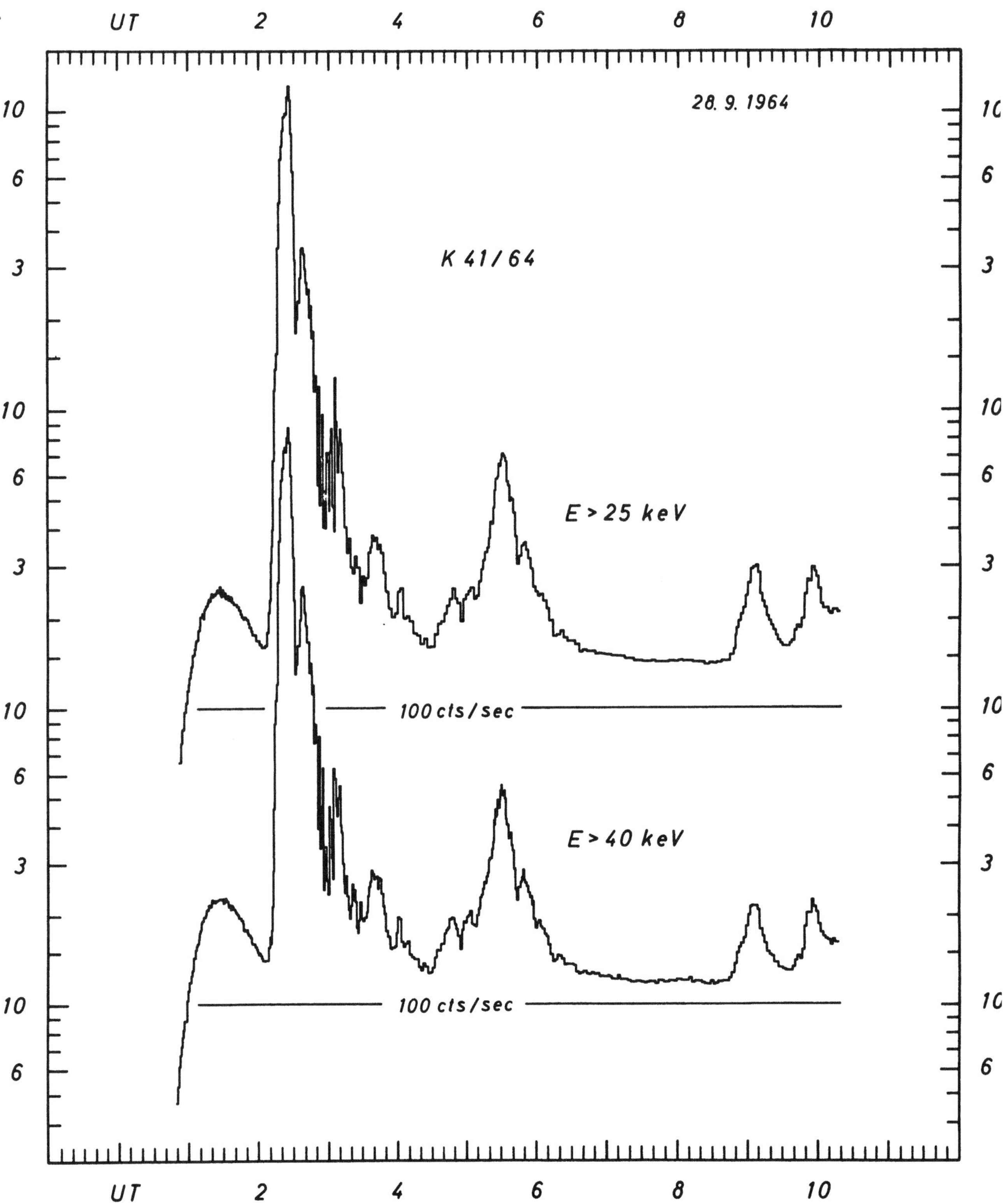

UT
2
4
6
8
10
28. 9. 1964
10
6
3
K 41/64
10
6
3
E > 25 keV
100 cts/sec
10
6
3
E > 40 keV
100 cts/sec
10
6
UT
2
4
6
8
10

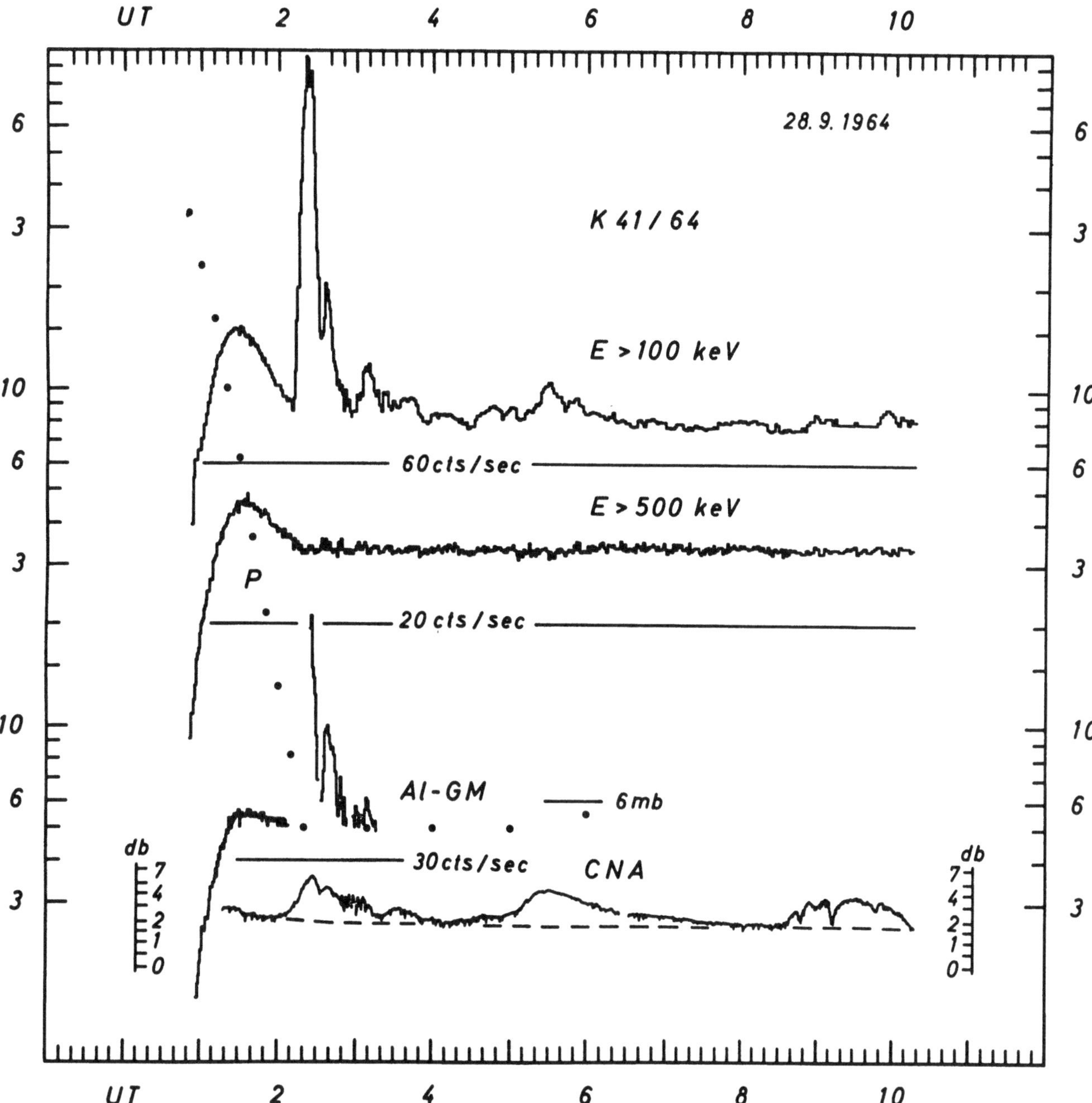

UT
2
4
6
8
10
28. 9. 1964
K 41 / 64
E > 100 keV
60 cts/sec
E > 500 keV
P
20 cts/sec
Al - GM
6 mb
30 cts/sec
CNA
db
7
4
2
1
0
UT
2
4
6
8
10

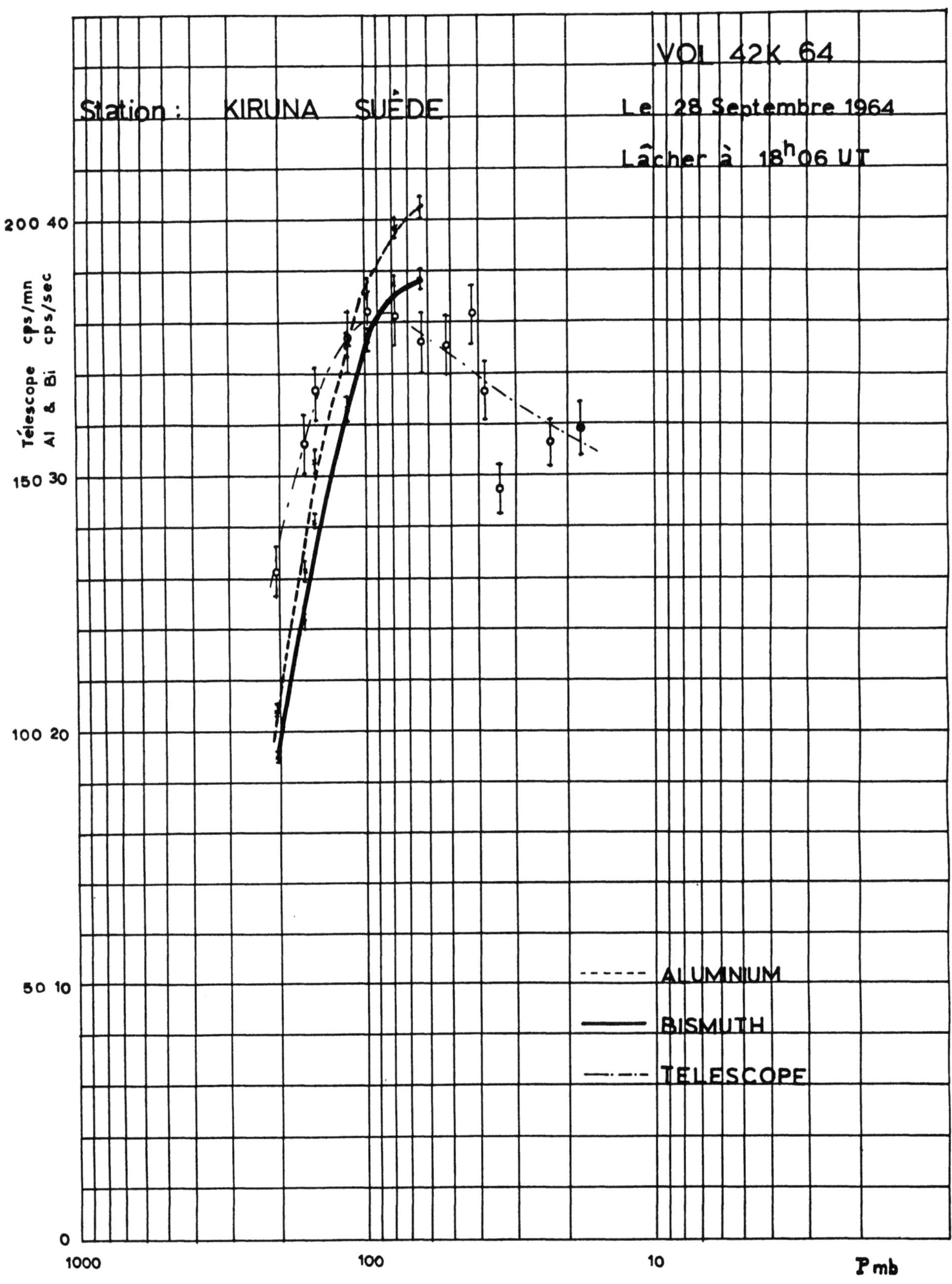

VOL 42K 64
Station :   KIRUNA   SUÈDE
Le 28 Septembre 1964
Lâcher à 18h06 UT
Télescope cps/mn
Al & Bi cps/sec
200 40
150 30
100 20
50 10
0
1000
100
10
P mb
ALUMINIUM
BISMUTH
TELESCOPE

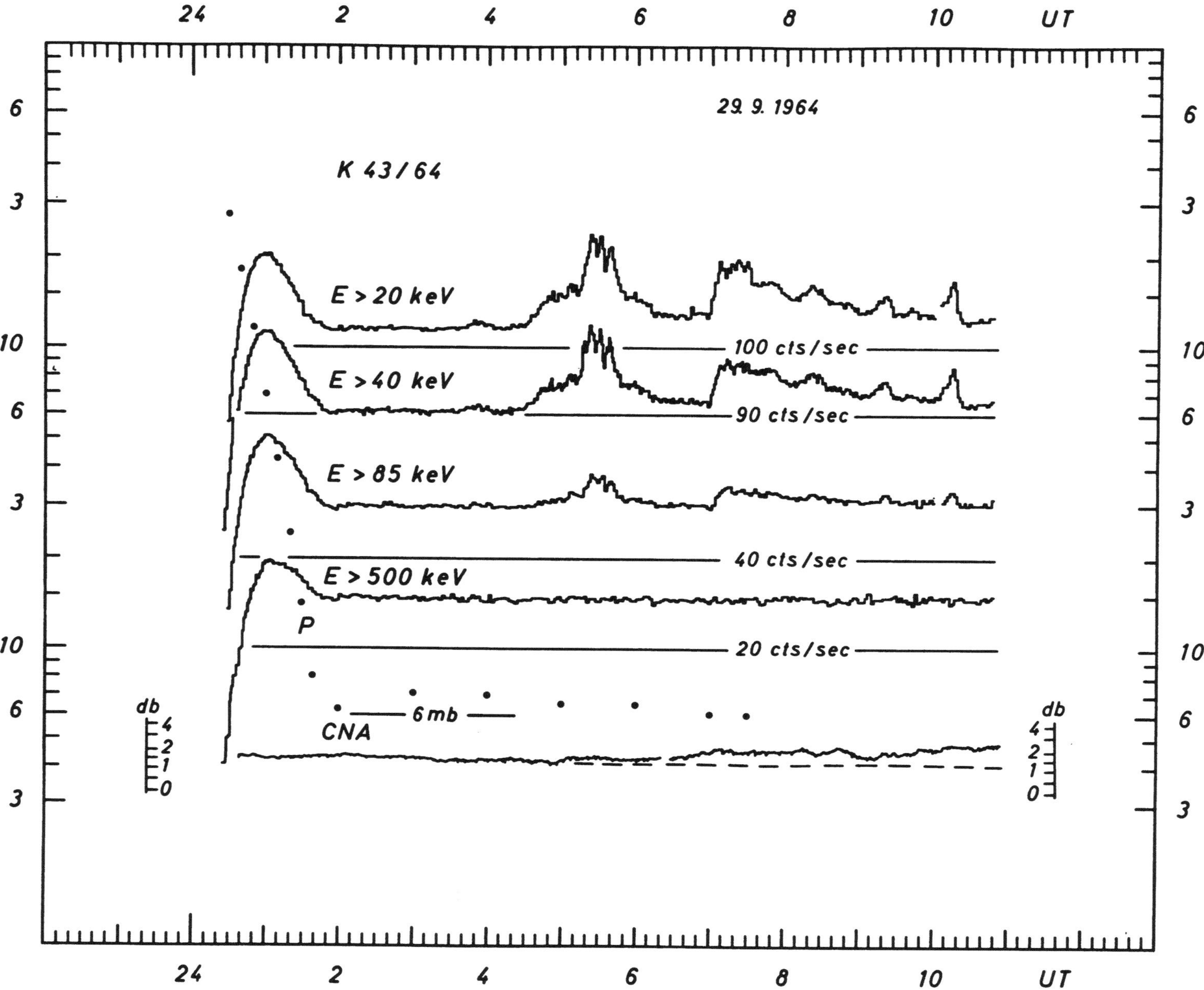

24
2
4
6
8
10
UT
29. 9. 1964
K 43/64
E > 20 keV
100 cts/sec
E > 40 keV
90 cts/sec
E > 85 keV
40 cts/sec
E > 500 keV
P
20 cts/sec
6 mb
CNA
db
4
2
1
0

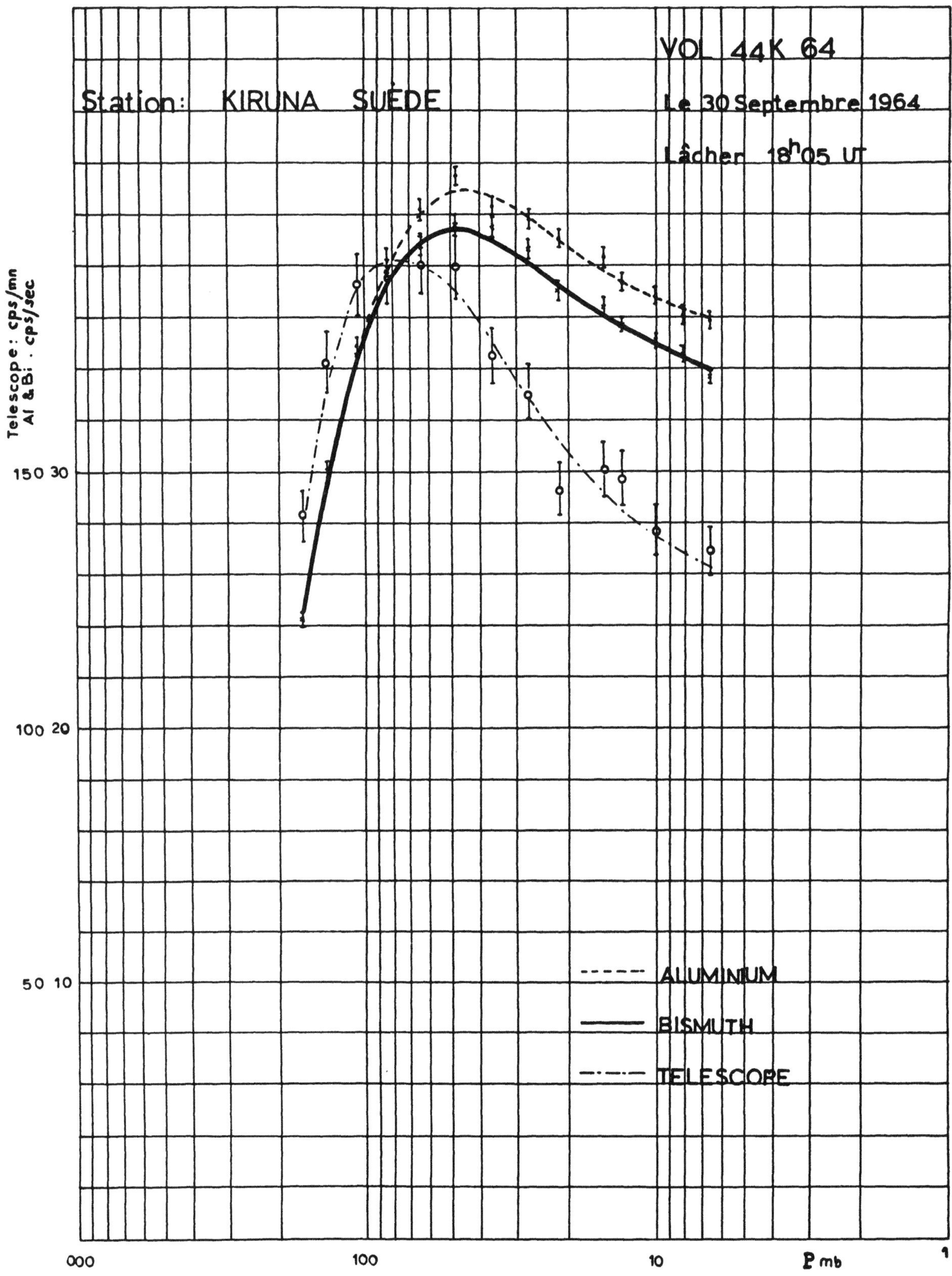

VOL 44K 64
Station: KIRUNA SUÈDE
Le 30 Septembre 1964
Lâcher 18h05 UT
Telescope : cps/mn
Al & Bi : cps/sec
150 30
100 20
50 10
ALUMINIUM
BISMUTH
TELESCOPE
000
100
10
P mb

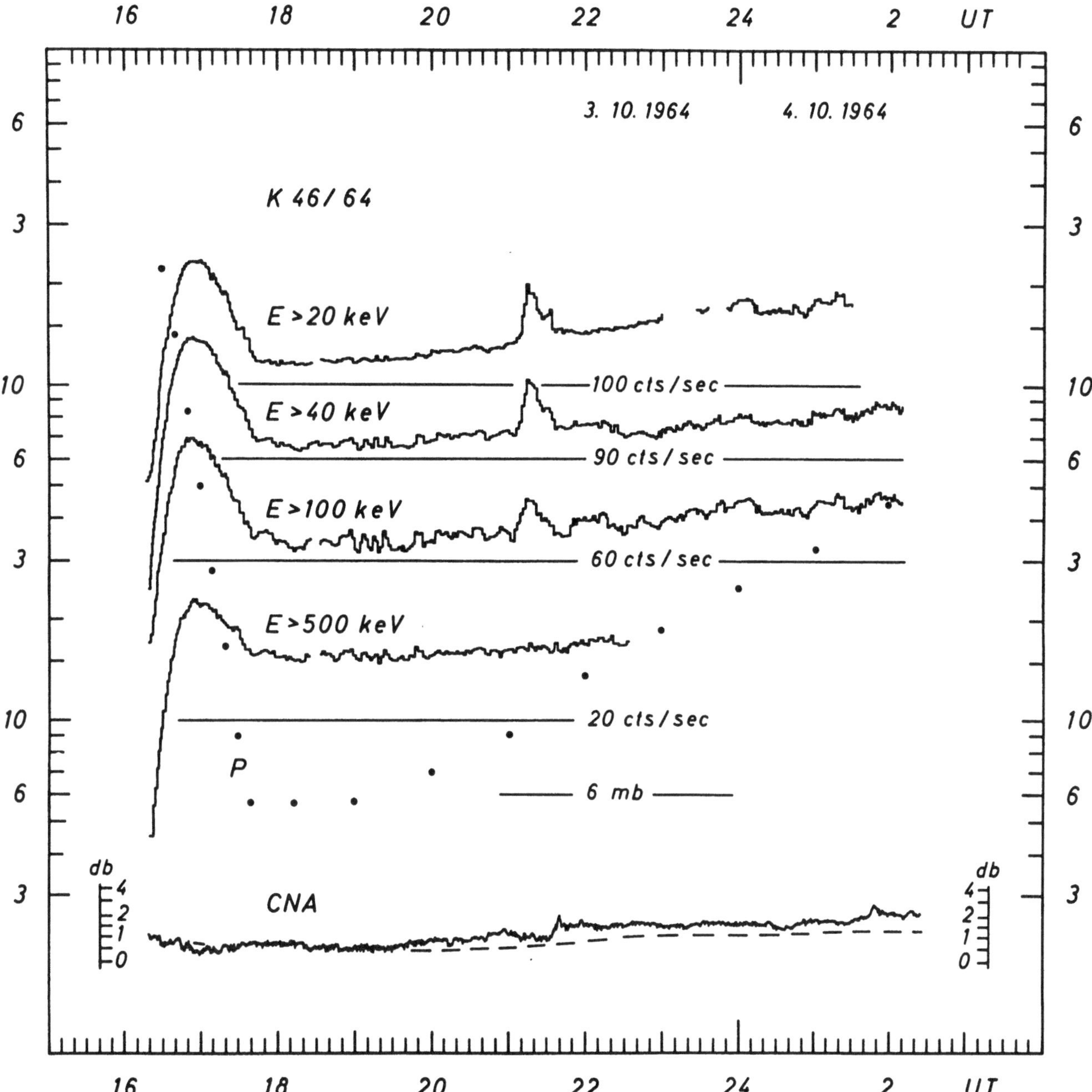

16   18   20   22   24   2   UT
3. 10. 1964   4. 10. 1964
K 46 / 64
E > 20 keV
E > 40 keV
E > 100 keV
E > 500 keV
100 cts / sec
90 cts / sec
60 cts / sec
20 cts / sec
6 mb
P
db
4
2
1
0
CNA
db
4
2
1
0
6
3
10
6
3
10
6
3
6
3
10
6
3
16   18   20   22   24   2   UT

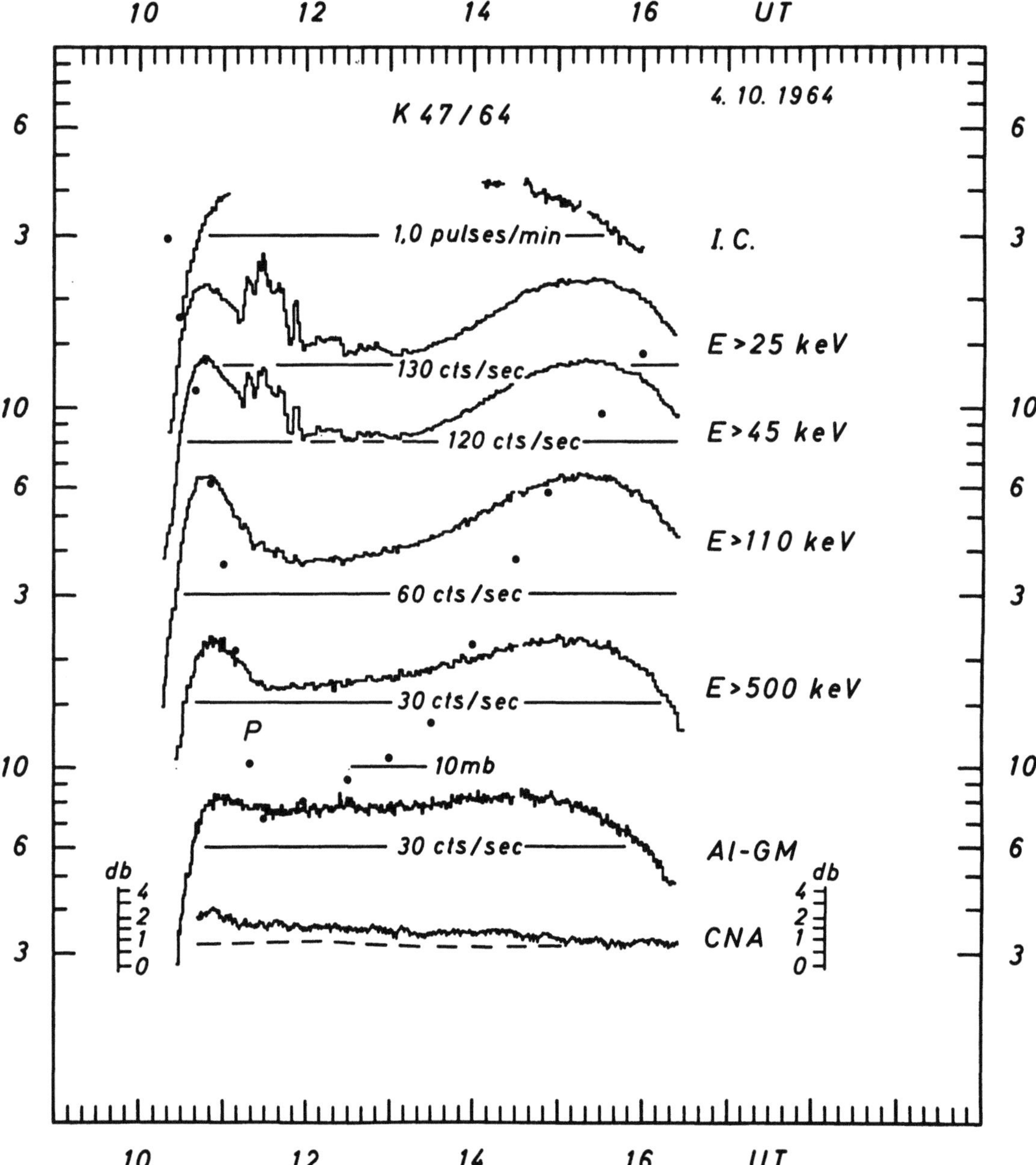

10
12
14
16
UT
K 47/64
4. 10. 1964
6
3
1,0 pulses/min
I. C.
E > 25 keV
130 cts/sec.
E > 45 keV
10
120 cts/sec
E > 110 keV
6
3
60 cts/sec
E > 500 keV
30 cts/sec
P
10
10mb
30 cts/sec
Al-GM
6
db
4
2
1
0
CNA
db
4
2
1
0
3
10
12
14
16
UT

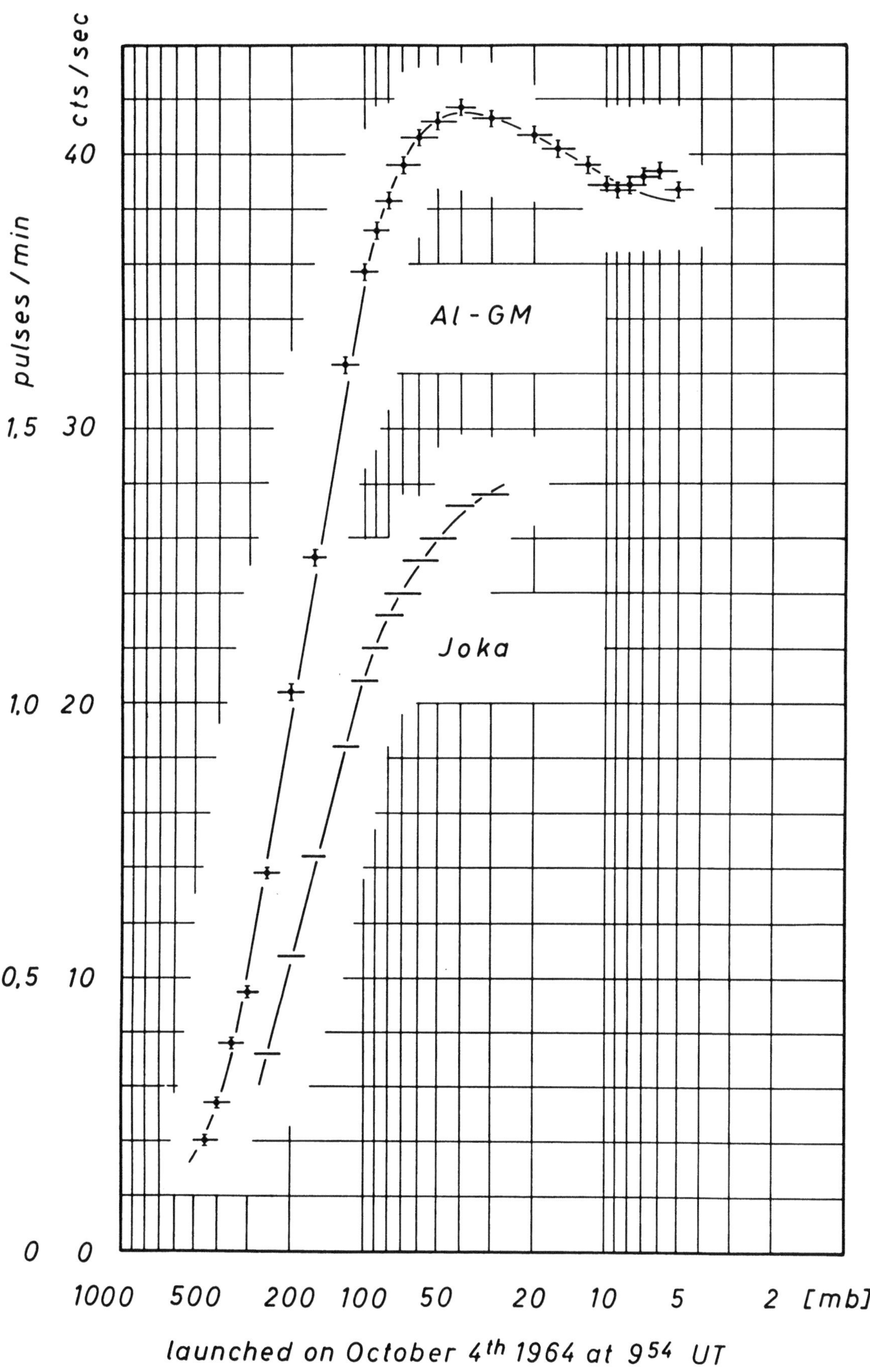

K 47/64
cts/sec
pulses/min
40
30
20
10
1,5
1,0
0,5
0
0
Al - GM
Joka
1000 500 200 100 50 20 10 5 2 [mb]
launched on October 4th 1964 at 9 54 UT

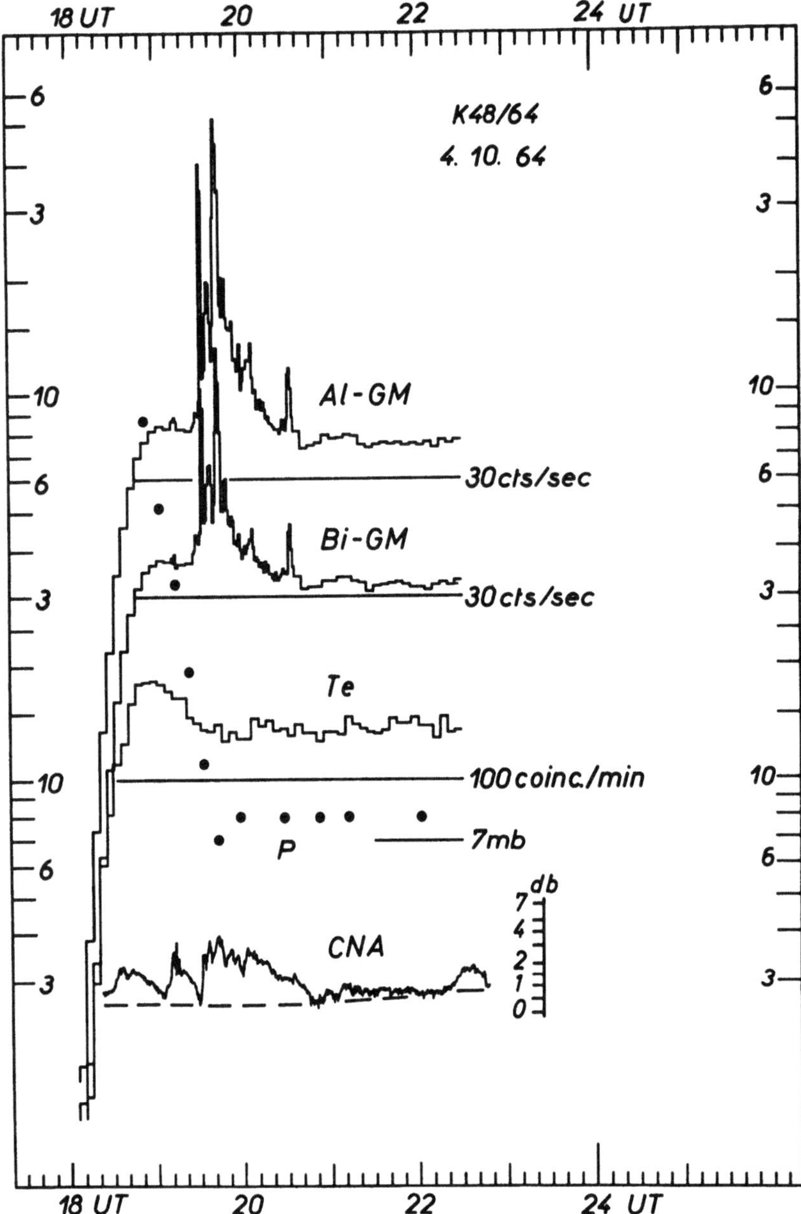

18 UT
20
22
24 UT
6
3
10
6
3
10
6
3
K48/64
4. 10. 64
Al-GM
30 cts/sec
Bi-GM
30 cts/sec
Te
100 coinc./min
P
7mb
db
7
4
2
1
0
CNA
6
3
10
6
3
10
6
3
18 UT
20
22
24 UT

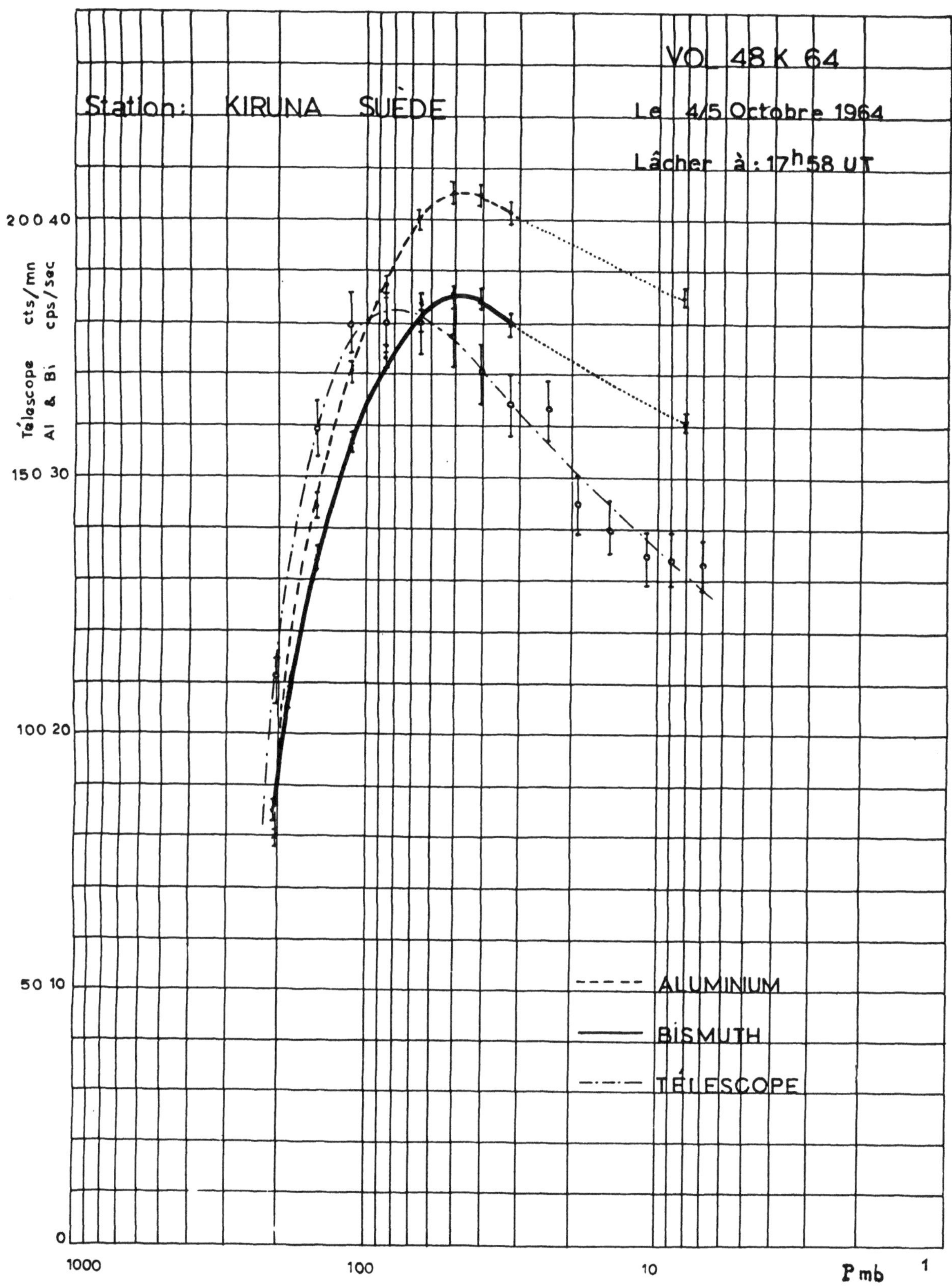

VOL 48 K 64
Station: KIRUNA SUÈDE
Le 4/5 Octobre 1964
Lâcher à: 17h58 UT
Télescope cts/mn
Al & Bi cps/sec
200 40
150 30
100 20
50 10
0
1000
100
10
P mb
1
ALUMINIUM
BISMUTH
TÉLESCOPE

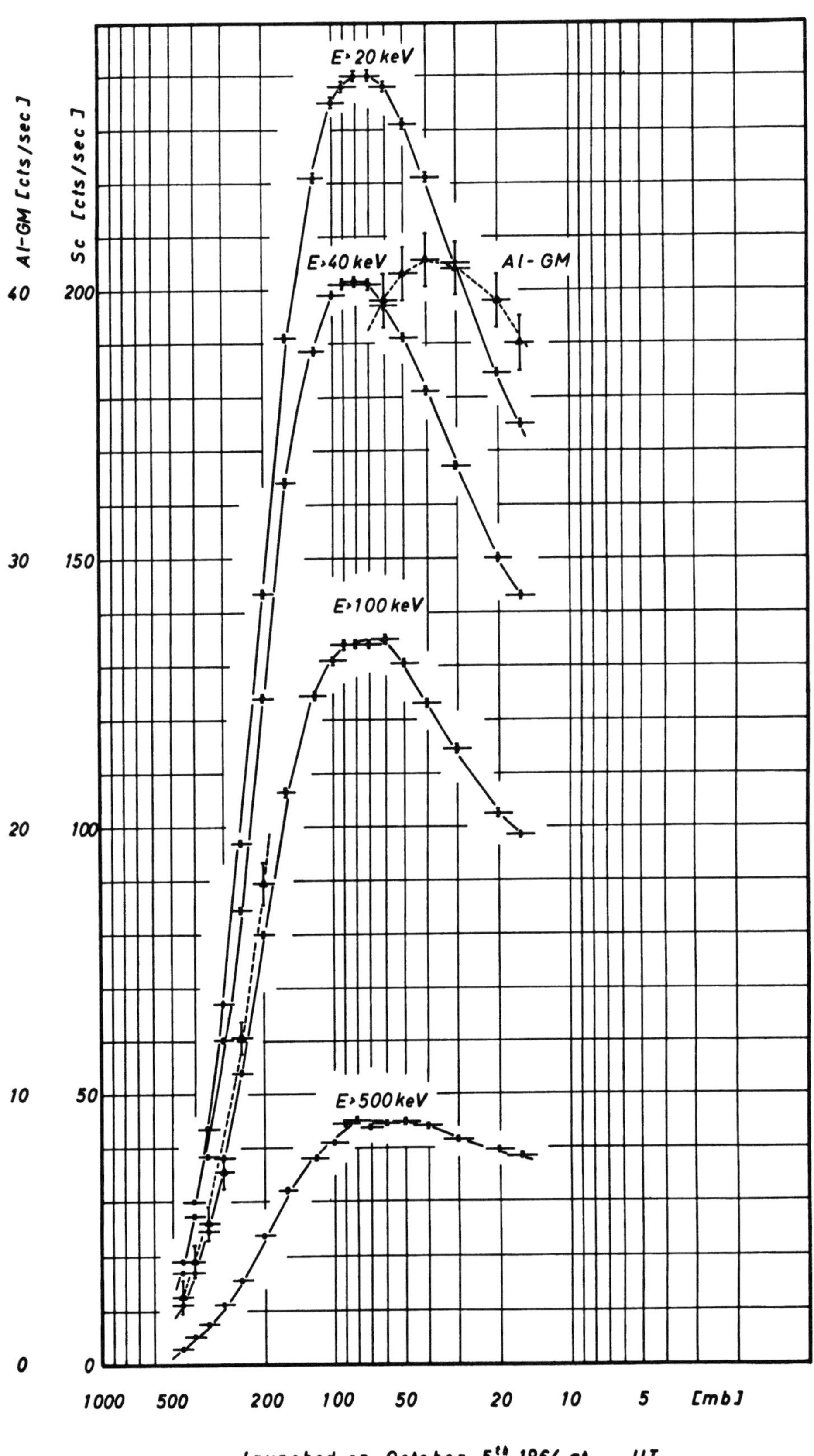

K 49 / 64
Al - GM [cts/sec]
Sc [cts/sec]
E › 20 keV
E › 40 keV
Al - GM
E › 100 keV
E › 500 keV
40
30
20
10
0
200
150
100
50
0
1000 500 200 100 50 20 10 5 [mb]
launched on October 5th 1964 at   UT

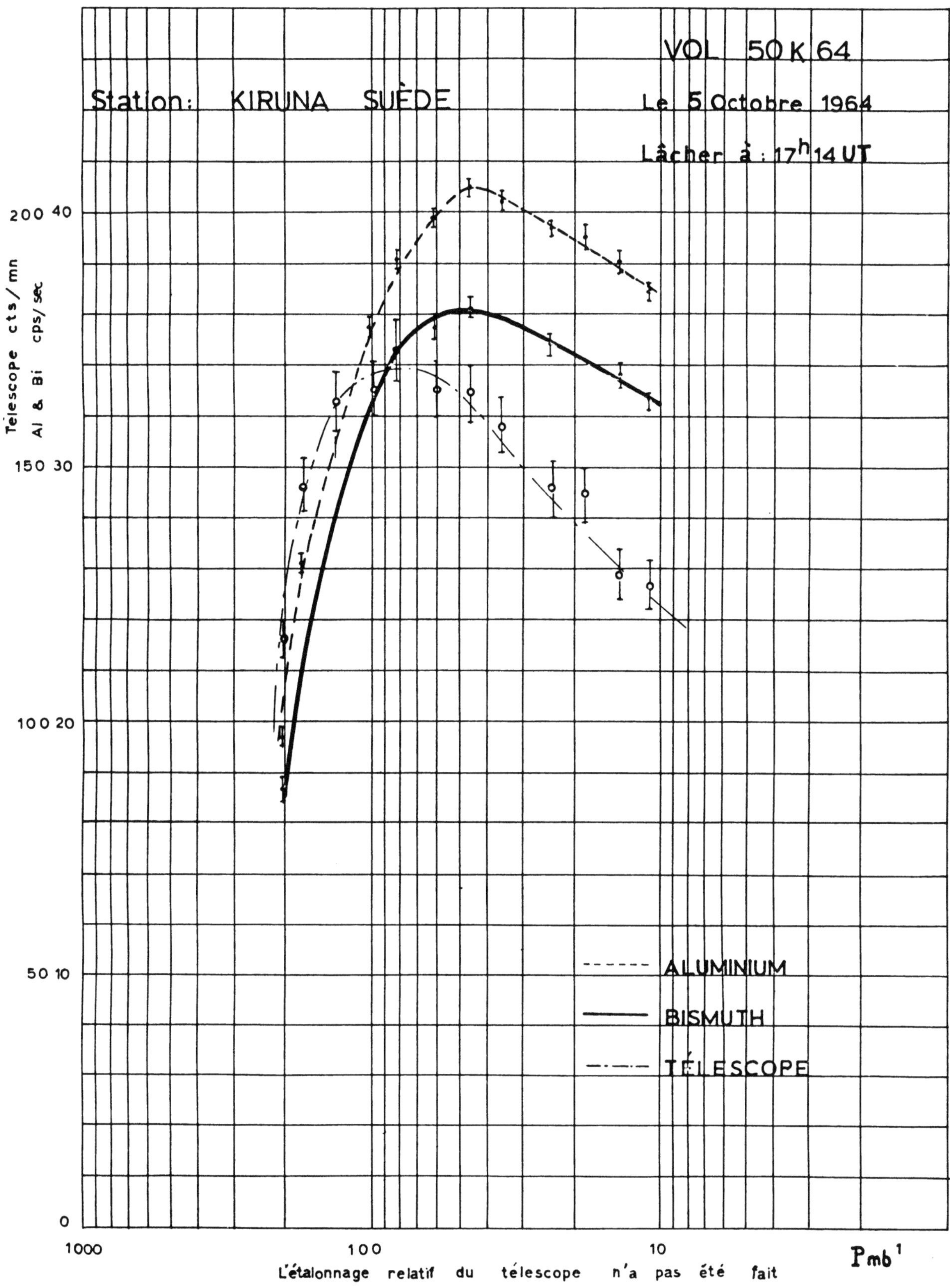

VOL 50 K 64
Station: KIRUNA SUÈDE
Le 5 Octobre 1964
Lâcher à : 17h 14 UT
Télescope cts / mn
Al & Bi   cps/sec
200 40
150 30
100 20
50 10
0
1000
100
10
Pmb
ALUMINIUM
BISMUTH
TÉLESCOPE
L'étalonnage relatif du télescope n'a pas été fait

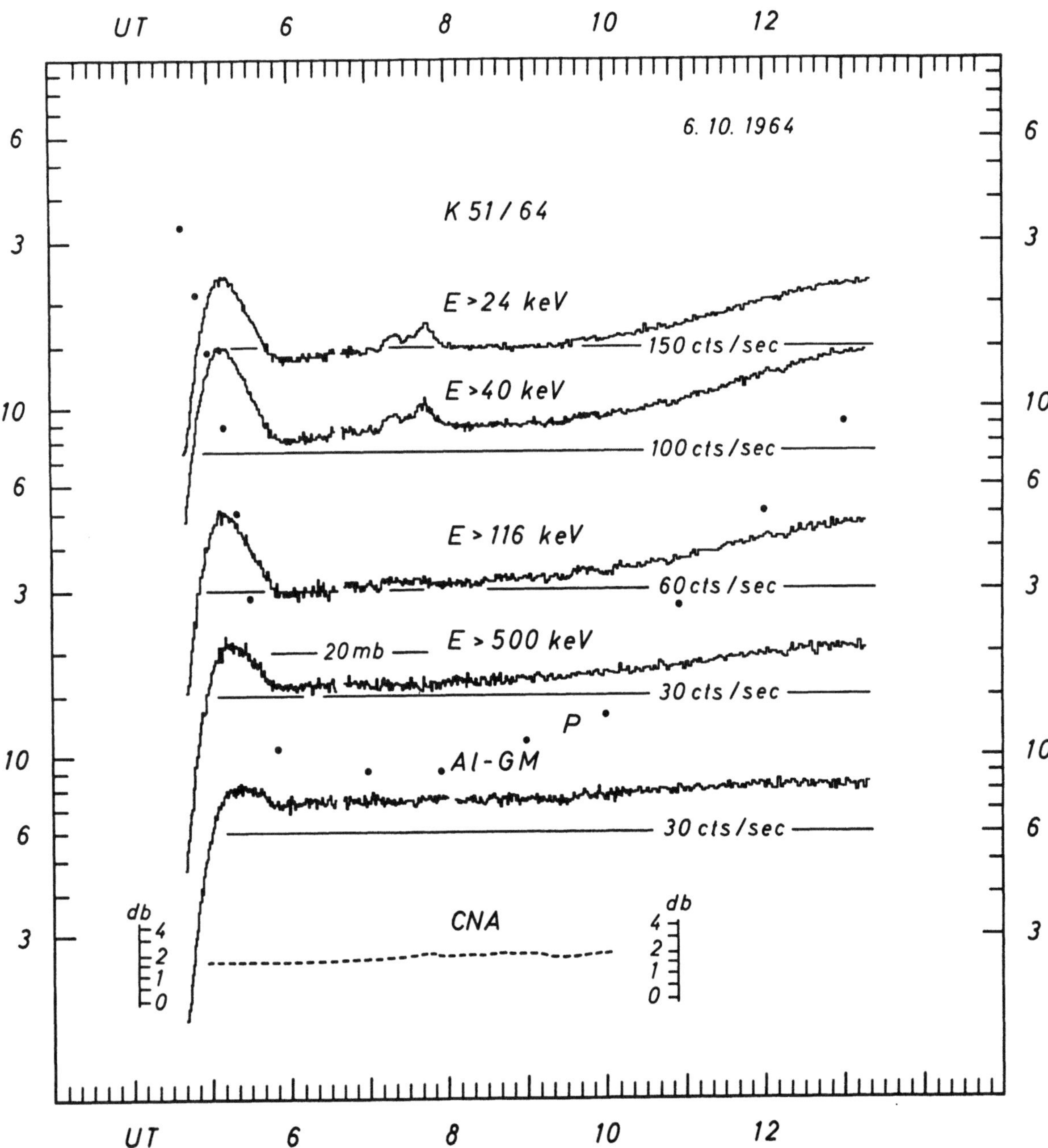

UT
6
8
10
12
6. 10. 1964
K 51 / 64
E > 24 keV
150 cts / sec
E > 40 keV
100 cts / sec
E > 116 keV
60 cts / sec
20 mb
E > 500 keV
30 cts / sec
P
Al - GM
30 cts / sec
db
4
2
1
0
CNA
db
4
2
1
0
UT
6
8
10
12

K 51/64

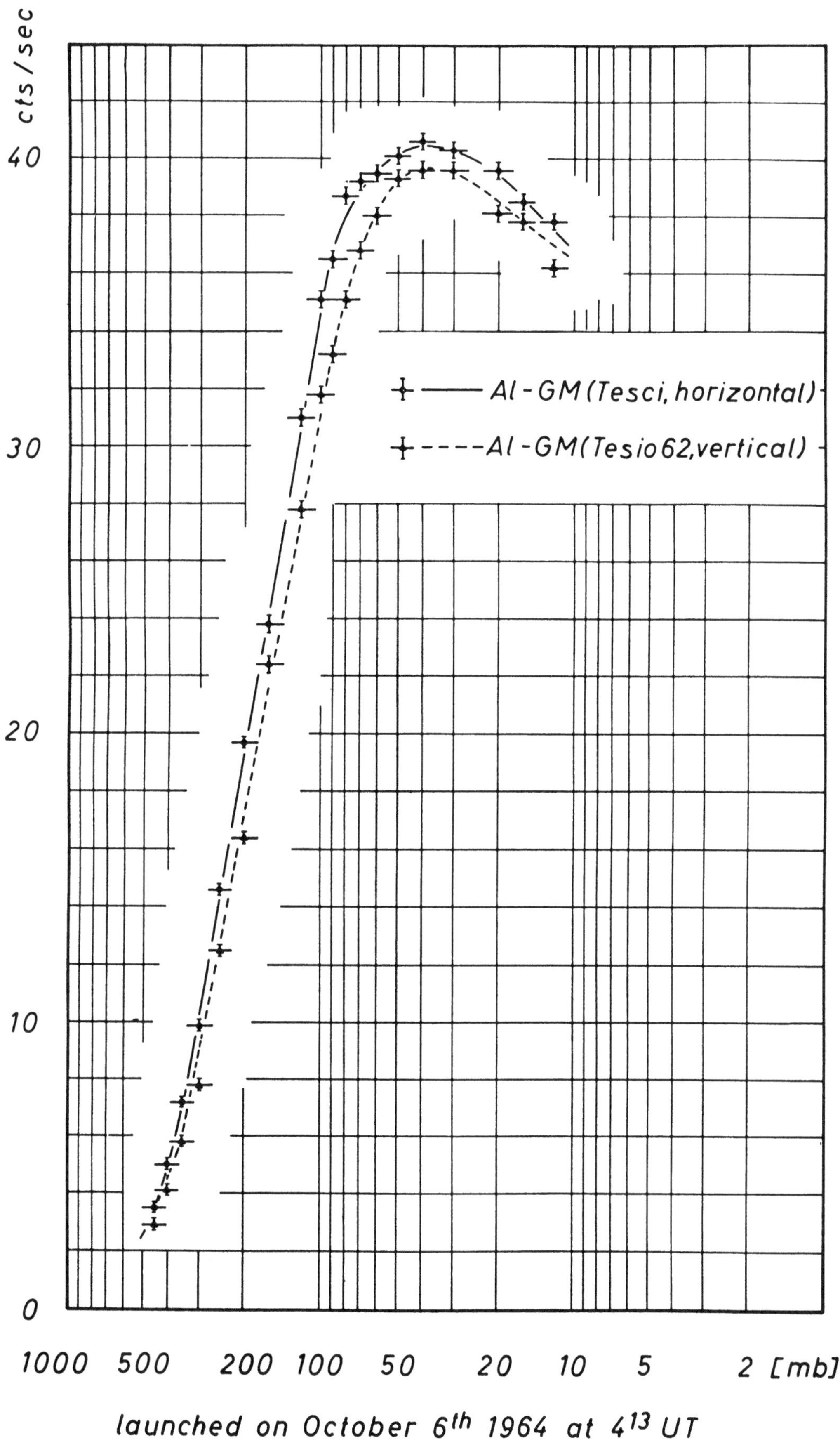

launched on October 6th 1964 at 4^{13} UT

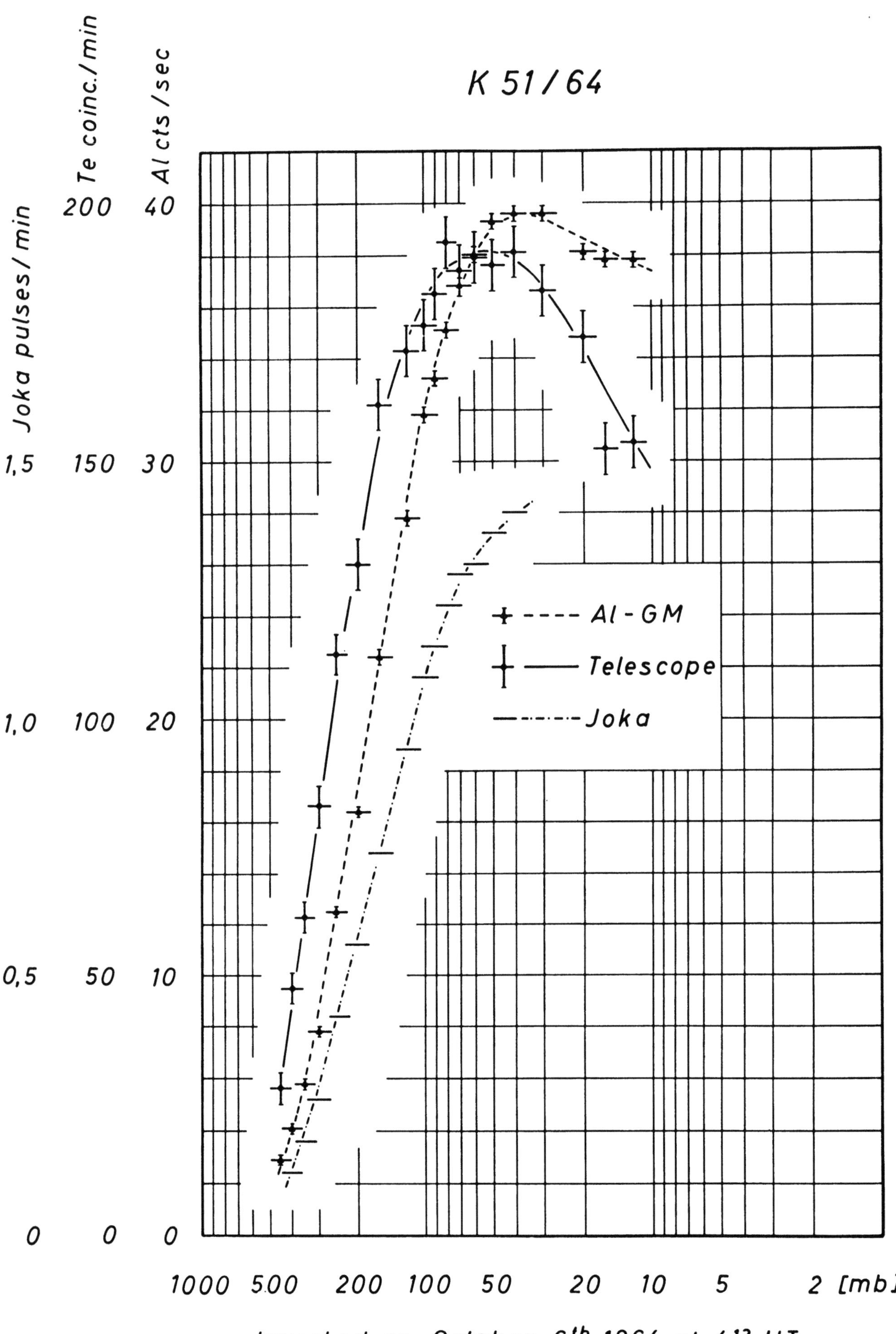

Te coinc./min
Al cts/sec
Joka pulses/min
K 51/64
200
40
150
30
100
20
50
10
1,5
1,0
0,5
0
0
0
Al - GM
Telescope
Joka
1000 500 200 100 50 20 10 5 2 [mb]
launched on October 6th 1964 at 4 13 UT

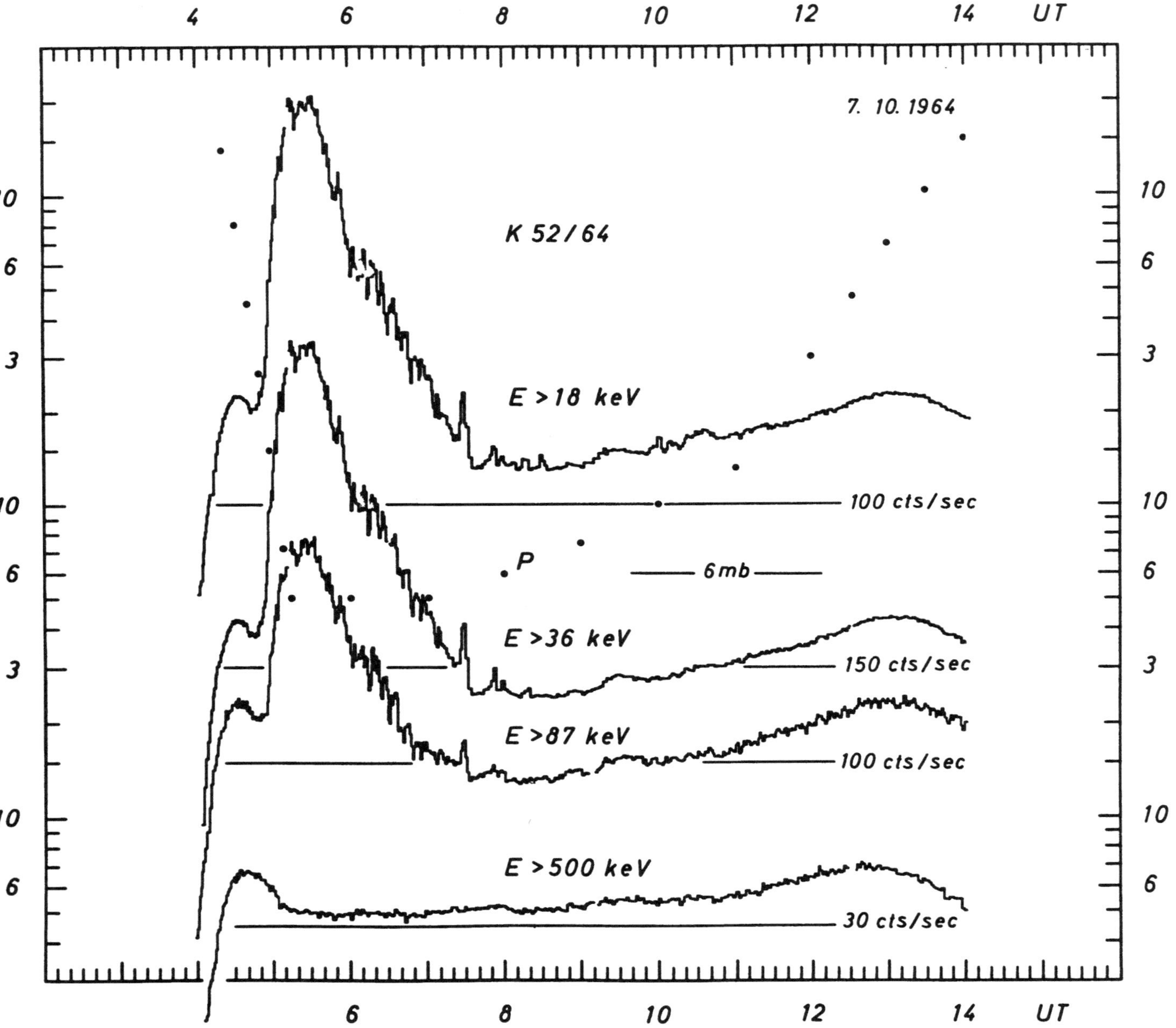

7. 10. 1964
K 52/64
E >18 keV
100 cts/sec
P
6mb
E >36 keV
150 cts/sec
E >87 keV
100 cts/sec
E >500 keV
30 cts/sec
10
6
3
10
6
3
10
6
4
6
8
10
12
14 UT

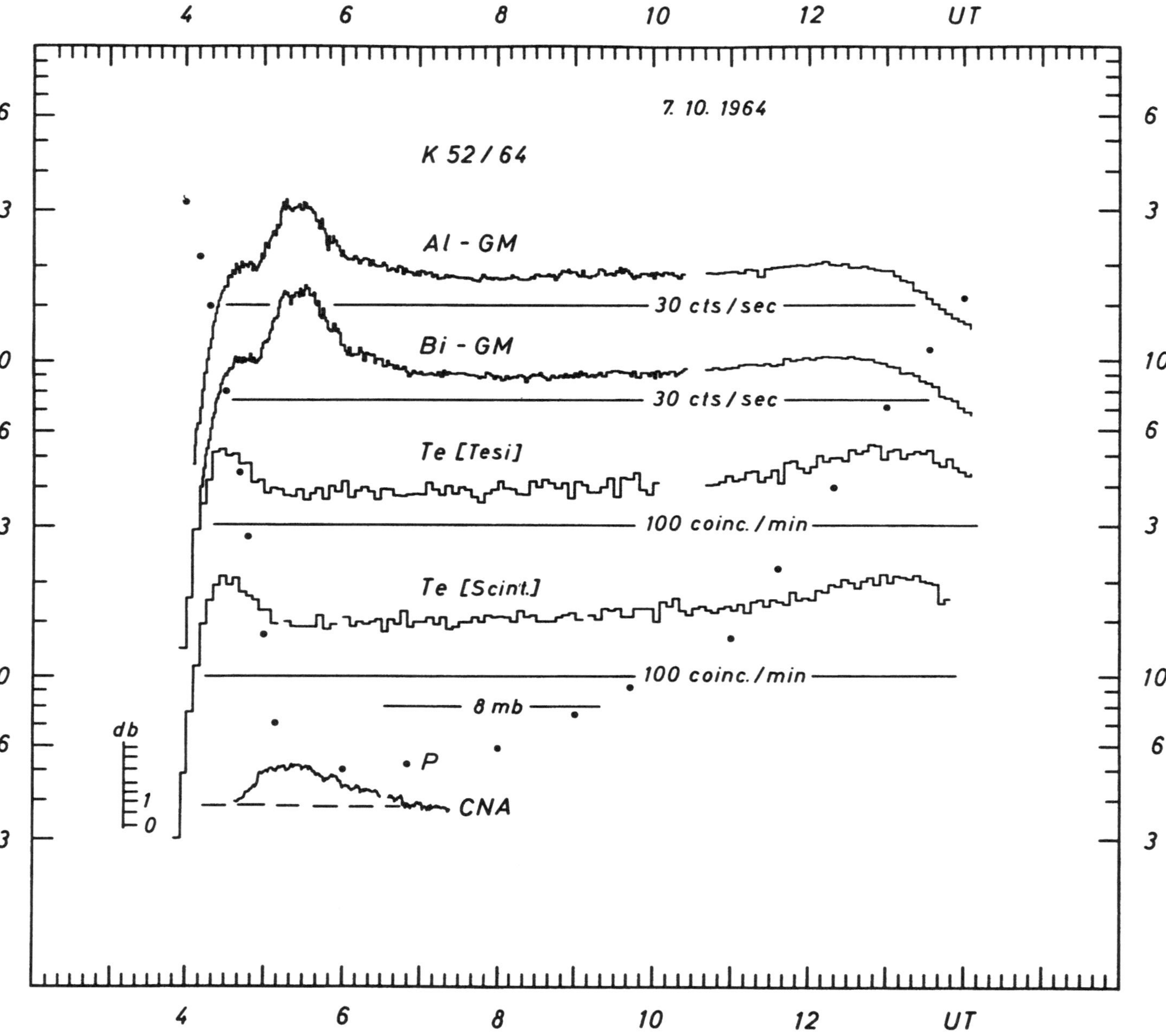

7. 10. 1964
K 52 / 64
Al - GM
30 cts / sec
Bi - GM
30 cts / sec
Te [Tesi]
100 coinc. / min
Te [Scint.]
100 coinc. / min
8 mb
db
1
0
P
CNA
UT
4
6
8
10
12
UT

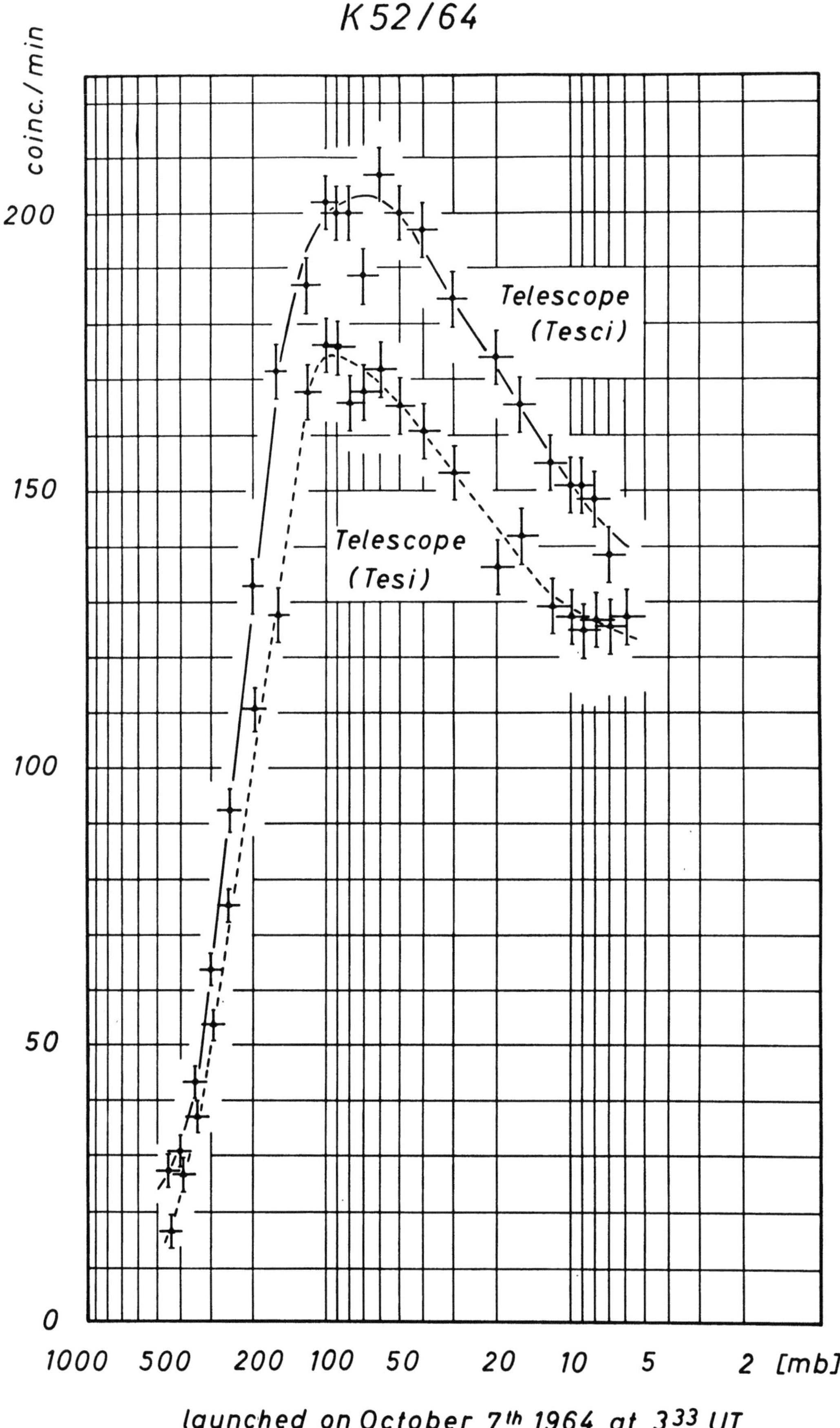

K 52/64
coinc./min
200
150
100
50
0
Telescope
(Tesci)
Telescope
(Tesi)
1000   500   200   100   50   20   10   5   2 [mb]
launched on October 7th 1964 at 3³³ UT

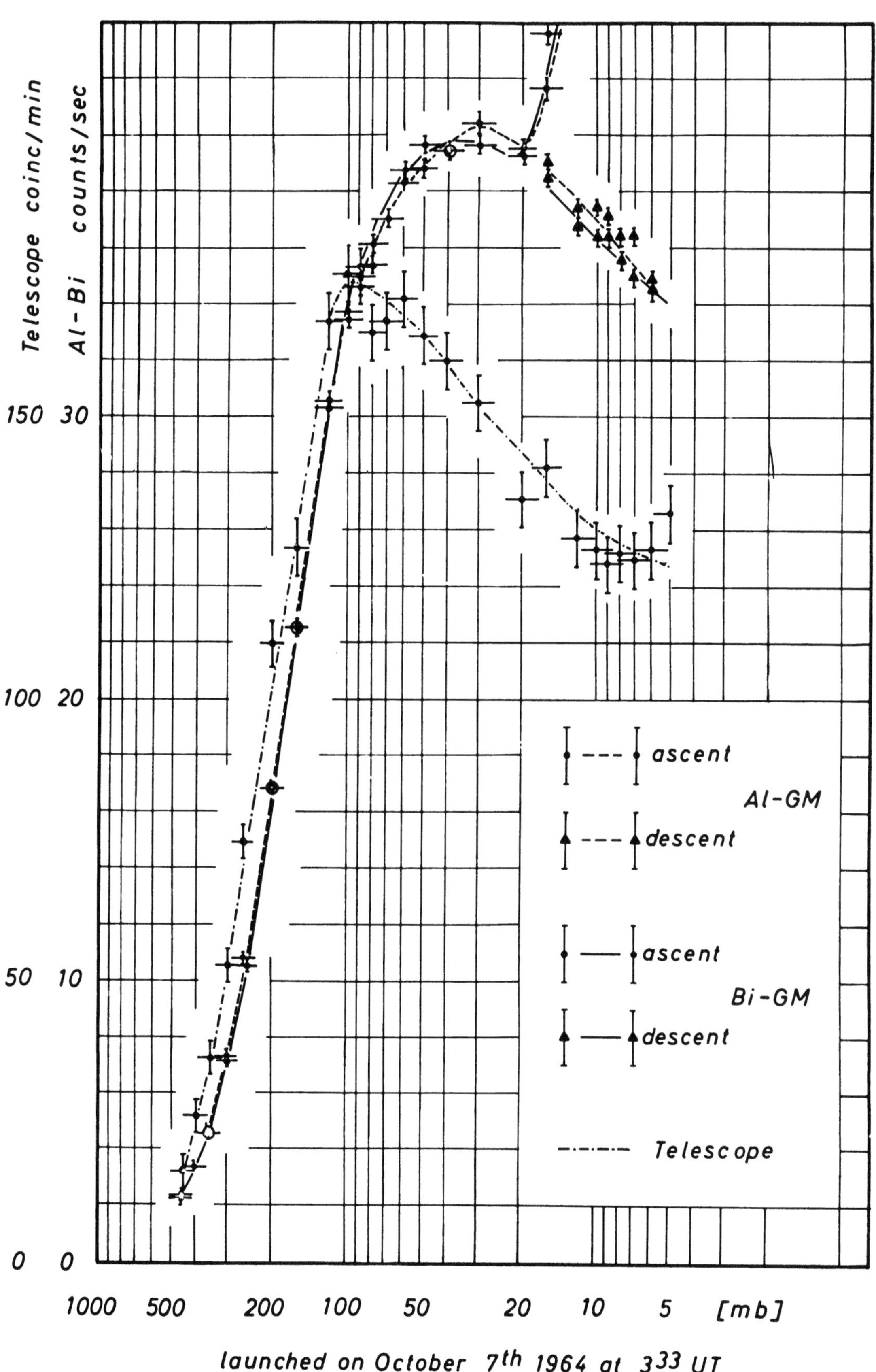

K 52 / 64
Telescope coinc/min
Al - Bi counts/sec
150 30
100 20
50 10
0 0
1000 500 200 100 50 20 10 5 [mb]
ascent
Al-GM
descent
ascent
Bi-GM
descent
Telescope
launched on October 7th 1964 at 3.33 UT

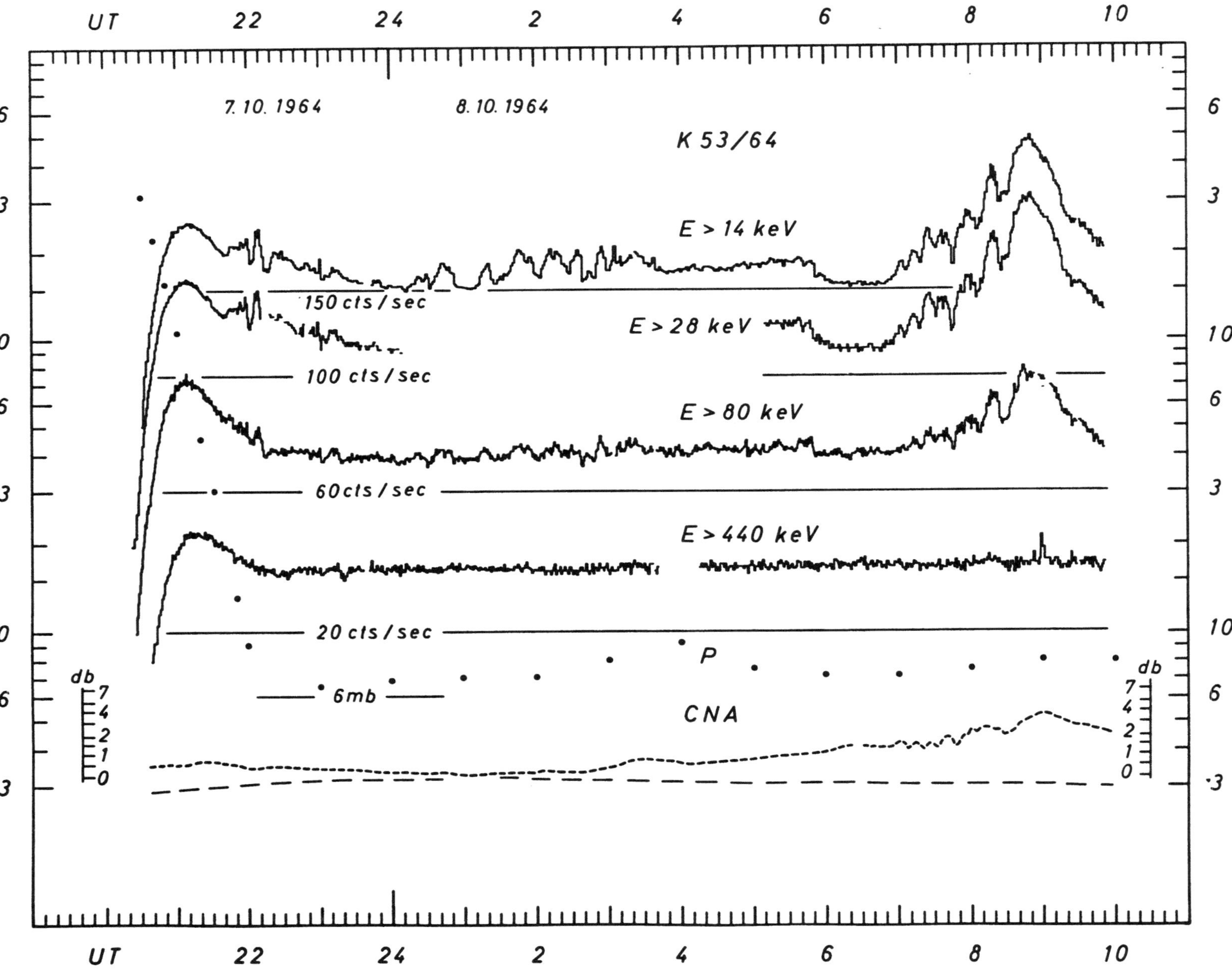

UT
7. 10. 1964
8. 10. 1964
K 53/64
E > 14 keV
150 cts/sec
E > 28 keV
100 cts/sec
E > 80 keV
60 cts/sec
E > 440 keV
20 cts/sec
P
db
7
4
2
1
0
6mb
CNA
UT

**Verzeichnis der Mitteilungen aus dem Max-Planck-Institut
für Physik der Stratosphäre**

Nr. 1/1953   Über den Beitrag der von μ - Mesonen angestoßenen Elektronen zu den Ultrastrahlungsschauern unter Blei. G. Pfotzer

Nr. 2/1954   Ein Zählrohrkoinzidenzgerät zur Registrierung der kosmischen Ultrastrahlung. A. Ehmert

Eine einfache Me hode zur Einstellung und Fixierung des Expansionsverhältnisses von Jebelkammern. G. Pfotzer

Nr. 3/1954   Optische Interferenzen an dünnen, bei -190°C kondensierten Eisschichten. Erich Regener (vergriffen)

Nr. 4/1955   Über die Messung der Temperatur des atmosphärischen Ozons mit Hilfe der Huggins-Banden. H. Zschörner und H. K. Paetzold

Nr. 5/1956   Ein neuer Ausbruch solarer Ultrastrahlung am 23. Februar 1956. A. Ehmert und G. Pfotzer, vergriffen (erschienen Z. Naturforschung 11a, 322, 1956)

Nr. 6/1956   Das Abklingen der solaren Ultrastrahlung beim Ausbruch am 23. Februar 1956 und die geomagnetischen Einfallsbedingungen. A. Ehmert und G. Pfotzer

Nr. 7/1956   Die Impulsverteilung der solaren Ultrastrahlung in der Abklingphase des Strahlungseinbruches am 23. Februar 1956. G. Pfotzer

Nr. 8/1956   Die atmosphärischen Störungen und ihre Anwendung zur Untersuchung der unteren Ionosphäre. K. Revellio

Nr. 9/1956   Solare Ultrastrahlung als Sonde für das Magnetfeld der Erde in großer Entfernung. G. Pfotzer

*

Die vorstehenden Hefte können beim Max-Planck-Institut für Aeronomie,
3411 Lindau angefordert werden.

Mitteilungen aus dem Max-Planck-Institut für Aeronomie

Nr. 1 (S) Waibel: Messungen von Primärteilchen der kosmischen Strahlung.

Nr. 2 (S) Erbe: Auswirkung der Variationen der primären kosmischen Strahlung auf die Mesonen- und Nukleonenkomponente am Erdboden.

Nr. 3 (I) Kohl: Bewegung der F-Schicht der Ionosphäre bei erdmagnetischen Bai-Störungen.

Nr. 4 (I) Becker: Tables of ordinary and extraordinary refractive indices, group refractive indices and $h'_{o,x}(f)$-curves or standard ionospheric layer models.

Nr. 5 (S) Schröpl: Über eine Neubestimmung des Absorptionskoeffizienten von Ozon im Ultraviolett bei kleinen Konzentrationen.

Nr. 6 (S) Erbe: Ergebnisse der Ballonaufstiege zur Messung der kosmischen Strahlung in Weissenau und Lindau.

Nr. 7 (S) Meyer: Elektromagnetische Induktion eines vertikalen magnetischen Dipols über einem leitenden homogenen Halbraum.

Nr. 8 (I u. S) Dieminger und Mitarb.: Die geophysikalischen Ereignisse des 12. - 14. November 1960.

Nr. 9 (S) Pfotzer, Ehmert, and Keppler: Time Pattern of Ionizing Radiation in Balloon Altitudes in High Latitudes. Part A, Text; Part B, Figures and Diagrams.

Nr. 10 (S) Waibel: Eine Ballonsonde zur Messung von Röntgenstrahlung und solarer Ultrastrahlung.

Nr. 11 (S) Voelker: Zur Breitenabhängigkeit erdmagnetischer Pulsationen.

Nr. 12 (S) Jaeschke: Registrierung von Pulsationen im südlichen Niedersachsen als Beitrag zur erdmagnetischen Tiefensondierung.

Nr. 13 (S) Meyer: Elektromagnetische Induktion in einem leitenden homogenen Zylinder durch äußere magnetische und elektrische Wechsel- felder.

Nr. 14 (S) Kremser: Über den Zusammenhang zwischen Röntgenstrahlungs-Aus- brüchen in der Polarlichtzone und bayartigen erdmagnetischen Störungen.

Nr. 15 (S) Keppler: Messung von Röntgenstrahlung und solaren Protonen mit Ballongeräten in der Nordlichtzone.

Nr. 16 (S) Kirsch: Die Anisotropien der kosmischen Strahlung.

Nr. 17 (S) Guilino: Ausbau eines Wechsellichtmonochromators und seine Anwen- dung zur Messung des Luftleuchtens während der Dämmerung und in der Nacht.

Nr. 18 (S) Pfotzer and Ehmert: Measurements of High Energetic Auroral Radiations with Balloon - Borne Detectors in 1962 and 1963 Part A to C, Text; Part D, Figures and Diagrams.